HIDDEN TREASURES

NATURAL HISTORY MUSEUM ABU DHABI

HIDDEN TREASURES

NATURAL HISTORY MUSEUM ABU DHABI

متحف التاريخ الطبيعي أبوظبي
Natural History Museum
Abu Dhabi

SKIRA

FOREWORD

A breathtaking new chapter in humanity's story of life on Earth begins at The Natural History Museum Abu Dhabi (NHMAD). This world-class institution brings to life the epic tale of our universe, showcasing some of the most spellbinding natural wonders ever discovered.

For the first time, the extraordinary history of our planet's interconnected biodiversity—millions of distinct species of plants, animals, fungi, and microorganisms—can be seen through an Arabian lens. No other natural history museum offers such a rare and precious voyage, both inspiring and empowering visitors while underscoring the urgent need to better protect the intricate abundance of life that surrounds us.

Within an immersive journey through 13.8 billion years of natural history across the galleries, visitors can discover how the Arabian Peninsula forms a central part of the museum's narrative, highlighting the region's unique contribution to Earth's story.

The Natural History Museum Abu Dhabi reflects the UAE's deep commitment to science, sustainability, and education, amplifying the importance of understanding Earth's rich tapestry for human wellbeing and survival. It highlights the global efforts required to conserve our planet's unique biodiversity.

NHMAD is a gateway to scientific knowledge and a vibrant platform for dialogue. The museum plays an active role in advancing studies in palaeontology, zoology, marine biology, molecular science and more, contributing to international conversations around sustainability, innovation, and the challenges of our time.

Inside this book, readers will discover a small sample of the rarest and most significant specimens in the museum's collection, displayed within immersive galleries that ignite curiosity and passion for the natural world. This launch marks an ambitious new era in Abu Dhabi's cultural journey, continuing Saadiyat Island's transformation into a global centre for research, collaboration, and discovery.

Institutions like Zayed National Museum, Louvre Abu Dhabi, Guggenheim Abu Dhabi, and Natural History Museum Abu Dhabi are more than architectural icons, they invite our community and the world to engage with the universal language of art, science, and nature.

We are excited to witness the impact that NHMAD will make, using the ancient past to inspire the next generation to question, discover, and take part in shaping a more sustainable future. A reflection of the museum's collections and narrative, this book has fuelled my own curiosity and learning, and it is my hope that others will also feel a deeper connection to the boundless beauty and fragility of our shared world.

Mohamed Khalifa Al Mubarak
Chairman, Natural History Museum Abu Dhabi

The Saadiyat Cultural District is a cornerstone of Abu Dhabi's cultural strategy, designed to position the emirate as a global arts hub. This ambitious initiative supports economic diversification, boosts cultural tourism and nurtures a vibrant creative ecosystem—aligned with Abu Dhabi's vision of a knowledge-based society.

The district hosts world-renowned institutions such as the Louvre Abu Dhabi, the Zayed National Museum, and the forthcoming Guggenheim Abu Dhabi. It attracts millions annually and fosters cultural exchange through educational programmes, workshops, and exhibitions promoting cross-cultural understanding.

The Natural History Museum Abu Dhabi, a landmark within the Saadiyat Cultural District, is set to become a premier centre for scientific research, education, and cultural enrichment. It will illuminate the story of Earth's evolution and biodiversity through immersive exhibits and exceptional specimens, inspiring curiosity across all ages. By collaborating globally, the museum enhances Abu Dhabi's stature in scientific excellence and cultural exchange, while promoting sustainability and environmental awareness. As a major cultural attraction, it will strengthen tourism and contribute significantly to the district's vibrant mosaic of artistic and educational experiences.

The establishment of the Natural History Museum Abu Dhabi exemplifies the emirate's commitment to becoming a leading cultural destination. This visionary institution is poised to elevate Abu Dhabi on the global stage, blending scientific excellence with immersive experiences that inspire discovery and stewardship. This strategy will not only enhance the emirate's cultural landscape, but also support broader economic and educational objectives.

Rita Aoun-Abdo
Executive Director – Culture Sector,
Department of Culture and Tourism - Abu Dhabi

The natural world is full of wonders. In this volume we have selected some of the most spectacular specimens from our collections. Sometimes you will notice them right away in our galleries because of their size or their striking appearance. Others may be more difficult to spot at first glance—you might even miss them when walking through the museum. But a closer look reveals remarkable details; colours, patterns, joints, structures, things easily overlooked at first glance. The more attention you give these details, the more fascinating nature becomes.

With this book, we offer an opportunity to pause, study, and reflect. We invite you to enjoy the often-hidden features of plants, animals, rocks, minerals, and meteorites. Once you see them, you will notice such details in other places, not just at our museum, but in any natural history museum, and even on your next walk outdoors. The fine lines in a leaf; the mathematical simplicity of seashells; the pollen in the bristles of a bumblebee; the feet of a gecko; dewdrops in a spiderweb; animal tracks in sand; and the skilfully co-ordinated flight of migratory birds.

Natural history museums are special. Together, between all of us, we are taking care of our shared natural heritage. In our collections, you will find evidence of what we know about the natural world, how species have evolved, and how everything in nature is connected. Our collections are at the centre of all we do. We have a responsibility to safeguard, develop, and enable access to them. We want to share them with everyone—to let you see what we see, and perhaps maybe make new discoveries or find personal meanings that might inspire you to share your personal stories of nature with others.

I am always taken aback by the fact that every single specimen has a story. Some tell tales older than our Solar System and Earth itself. Others speak of lost worlds long gone: of snow and ice, volcanoes and earthquakes, of the deepest seas and highest mountains, of abundance and catastrophe, of survival and extinction. Many reveal how extraordinary and diverse life is, the incredible forms it takes, and how vital it is for us to care for our planet.

I hope this book will inspire you to take a closer look at all the amazing things in our museum and helps you to forge a deeper connection.

Dr Peter C. Kjærgaard
Director of the Natural History Museum Abu Dhabi

INTRODUCTION
HIDDEN TREASURES

Amid the awe-inspiring displays of towering dinosaurs and the vast skeleton of a blue whale, countless wonders lie quietly within the Natural History Museum Abu Dhabi. Each object in this remarkable collection tells a story, some are obvious while others are more subtle, but many risk being overlooked when overshadowed by giants. *Hidden Treasures* invites readers to pause, look closer, and rediscover the beauty concealed within these 'endless forms most beautiful' that together narrate our planet's 4.5-billion-year journey.

This volume marks the first in a series celebrating a natural history museum collection unlike any other in the region. From a speck of stardust to the bones of titanic sauropods, each specimen reveals a fragment of the extraordinary story of our Universe. This extraordinary journey began 13.8 billion years ago, at the dawn of time, and gradually unfolded to nurture the emergence of life on Earth. The Murchison Meteorite, containing seven-billion-year-old stardust forged in ancient stars, connects us to our cosmic origins. If you have ever dreamt of reaching up to the stars, you are already among them. Rocks on our planet dating back 3.7 billion years provide the earliest tangible evidence of life's evolution, while chemical clues trace back even further—4.1-billion-years—hinting at the earliest spark of biology. These treasures not only connect us to the deep past, but also remind us of the delicate balance between planetary systems and life itself.

As life evolved through the ages, it blossomed into the rich biodiversity we see today, yet this foundation is now threatened by human activity. *Hidden Treasures* invites reflection on extinction and survival in what many regard as Earth's sixth mass extinction event. While the world's attention often turns to charismatic megafauna, countless other species face quiet peril, invisible to most eyes, yet critical to our planet's health. All life is precious, intricately connected across time and space, like an ancient library, where the loss of a single page sends tremors through the rest of the living world.

Through the pages of this book, we share images of the objects that embody our mission: to inspire curiosity, creativity, and care for the natural world. We aim to become a world leader in interpreting and engaging with the natural heritage of Abu Dhabi, the United Arab Emitares, and the wider world, while also walking a responsible path toward a sustainable future. By exploring these hidden treasures, we invite readers to see not only the beauty of nature's past, but also the promise of our shared future. A future shaped by knowledge, wonder, and a renewed connection to the world we all call home.

Professor Phillip Lars Manning,
BSc(Hons), MSc, PhD, FGS, FRGS, FLS.
Director of Science, Natural History Museum Abu Dhabi

Ammonite

Schlotheimia sp.

Rare *Schlotheimia* ammonite from Monmouth Beach, Dorset, revealing intricate shell detail and offering a captivating glimpse into the ancient Jurassic seas of southern England, over 190 million years ago.

Ammonite

Microderoceras birchi

Preserved within a durable carbonate concretion, this 195-million-year-old *Microderoceras birchi* ammonite reveals exceptional detail from the Jurassic seas of southern England. The species is notable for its tightly coiled form and fine rib patterns, characteristics that make it an easily recognisable and valuable fossil.

Chinese mitten crab
Eriocheir sinensis

World-class travellers, Chinese mitten crabs live in freshwater but migrate to saltwater to reproduce, traveling over 1,400 kilometres upstream. Over several generations. They can even walk on land, appearing in streets, houses, pools, and airports worldwide.

Burning nettle
Urtica urens

The burning nettle earns its name from the fine, stinging hairs that blanket its lower leaves and stem, delivering a sharp, tingling sting on contact. This herbaceous annual has a delicate appearance that belies its potent defence.

Arabian green bee-eater
Merops cyanophrys

Arabian green bee-eaters nest in sandy banks, often in colonies, where relatives help feed chicks—a rare cooperative behaviour that reflects their strong social bonds and boosts chick survival.

Béréba eucrite meteorite

Falling in Burkina Faso in 1924, this meteorite slice from asteroid Vesta reveals a striking mix of minerals, resembling Earth's volcanic rock in colour and texture. This 900 g specimen is one of only three in existence.

Ensisheim ordinary chondrite

In 1492, a massive meteorite struck a wheat field in Alsace, France. Seen as a sign of divine favour, Austrian King Maximilian I had it chained and preserved in Ensisheim's church.

Carbo iron meteorite

Found in 1923 by cowboys in Sonora, Mexico, this 4.76-kilograms Carbo meteorite slice, cut from a 450-kilogram mass, reveals a crystalline texture formed as molten metal cooled deep within an asteroid.

Cape York iron meteorite

For centuries, Inuit people of northwest Greenland used Cape York meteorite fragments as 'great irons' for tools. This 5.4-kilogram cube reveals its rare crystalline structure, known as a Widmanstätten pattern, unique to iron meteorites.

European rose chafer

Cetonia aurata

European rose chafers show dazzling colour variation.
Adults thrive on flowers like roses, while larvae grow underground.
While this species occurs just in the museum collections in the UAE,
other chafer species can be found in UAE habitats.

Bonelli's eagle with a grey francolin
Aquila fasciata with *Ortygornis pondicerianus*

During breeding season, Bonelli's eagles loop, dive, and lock talons mid-air. Once heavily threatened by power line electrocutions, new mitigation technology has drastically reduced the number of deaths, aiding their survival.

Spiny plumeless thistle
Carduus acanthoides

The spiny plumeless thistle can produce over a thousand seeds per plant,
ensuring the survival of its offspring through prolific seed dispersal.
This reproductive strategy helps the species
persist and spread effectively.

Everlasting pea
Lathyrus latifolius

The brightly coloured flowers of the everlasting pea attract butterflies with their nectar. Native to Europe, North America, and Australia, these vigorous vining plants bloom throughout summer, adding vibrant splashes of rose, pink, or white to their surroundings.

Sanagua fruit

Manicaria saccifera

Sanagua fruits hide smooth, variegated seeds beneath their rough shells. Found in warm, swampy areas, these palms yield a light, earthy drink long valued for its subtle flavor and medicinal qualities.

Malayan Tapir
Tapirus indicus

The Malayan tapir's bold black-and-white pattern acts as clever camouflage. Its white 'saddle' breaks up its shape in forest light, helping this large, gentle giant stay hidden from predators.

Fossilised shark
Hybodus raricostatus

Hybodus raricostatus was a Jurassic shark predator, thriving in ancient seas for millions of years, and showcasing early shark evolution and marine ecosystem diversity.

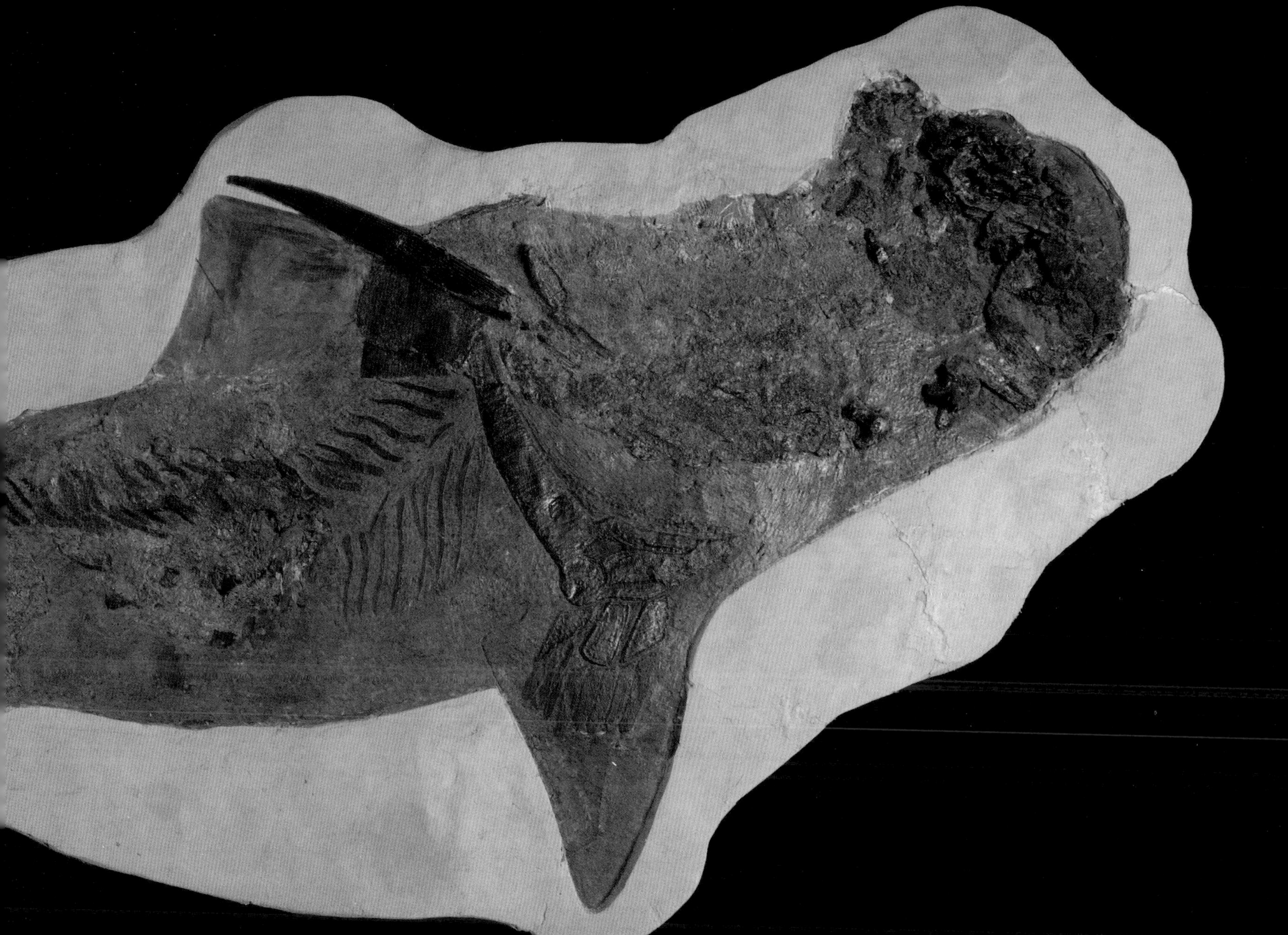

Western barn owl
Tyto alba

Barn owls are silent hunters and farmers' secret weapon, devouring up to 3,000 rodents a season. Their stealthy feasts cut chemical use, keeping soil rich and farmlands teeming with life.

Abu Dhabi
seven million years ago:

Crocodile skull (right)
Crocodylus sp.

Crocodile fossils found in Abu Dhabi reveal fierce apex predators once thrived in a lush river habitat, long before the region evolved into the arid desert it is today.

Following pages

Plesiogulo sp. (top left)

A fossilized lower jaw of a wolverine-like animal found in Abu Dhabi shows that this seven-million-year-old predator probably hunted a prehistoric ancestor of modern-day cane rats.

Crocodile jaw (bottom left)
Ikanogavialis sp.

This slender fossil jaw belongs to an extinct gavial-like crocodile. The teeth were particularly suited to devouring fish, which once lived in the rivers of Abu Dhabi. Similar looking crocodiles now inhabit rivers in India and Nepal.

Bovid species (top right)
Pachyportax latidens

Abu Dhabi's ancient savannahs were home to various species of cattle, bison, buffalo, and antelope. One of these was this giant, extinct bovine species—represented here by a fossil of its lower jaw.

Hipparion abudhabiense (bottom right)

Encased in resin, this molar from an extinct three-toed horse found in the Baynunah Formation was cross-sectioned and polished, offering scientists a rare glimpse into the region's prehistoric world.

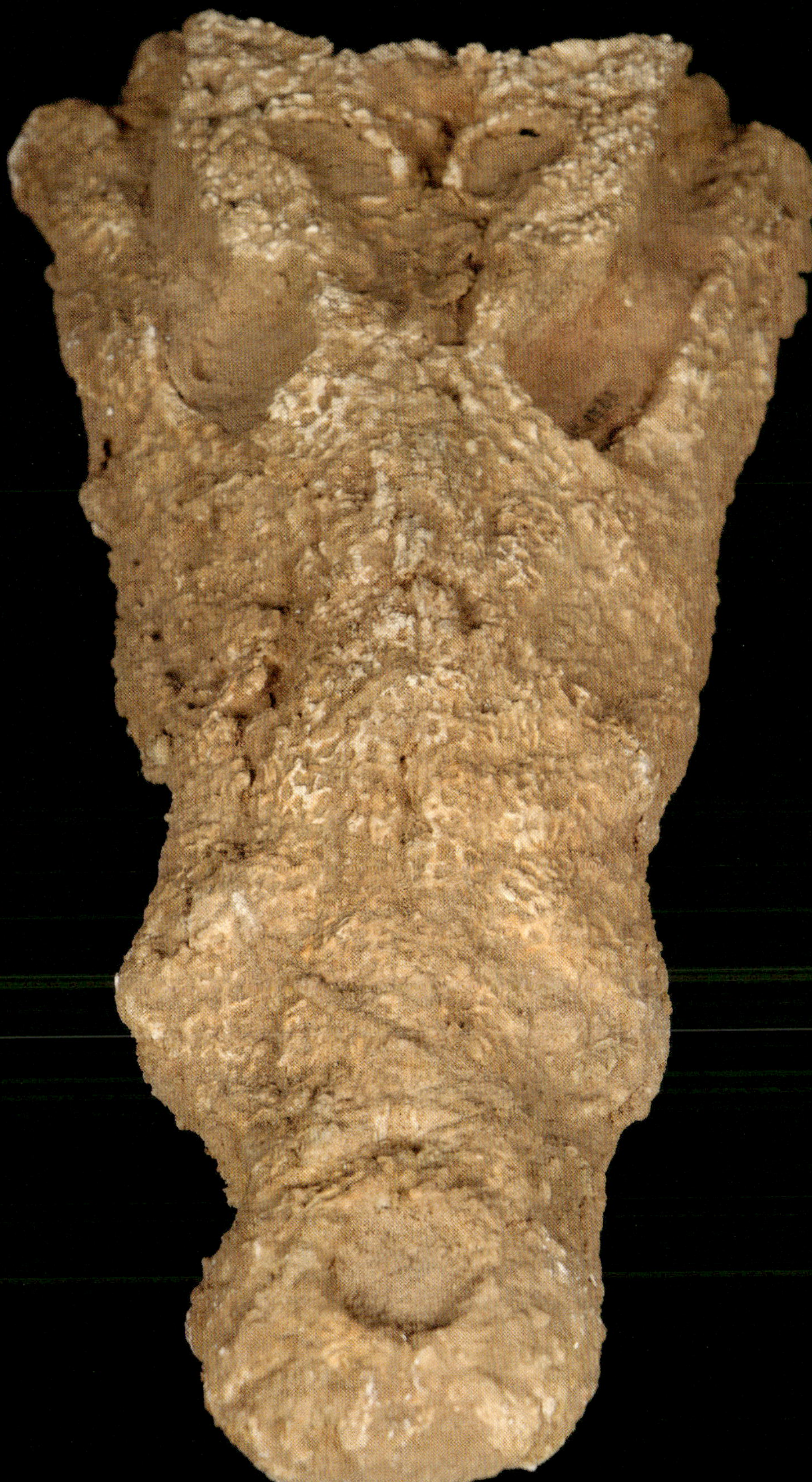

AUN 266

Western red panda
Ailurus fulgens

The western red panda's 'false thumb' isn't a real digit but a clever wrist bone extension. It grips bamboo with ease, mirroring the giant panda's design through stunning convergent evolution.

Arabian fat-tailed scorpion
Androctonus crassicauda

True to its name,
this medically significant scorpion roams across the Middle East,
including the UAE. Active at night and hidden by day,
it glows brilliantly under ultraviolet light.

Stag beetle

Lucanus cervus

Europe's largest beetle, the majestic male stag beetle, flaunts dramatic antler-like mandibles—perfect for fierce duels and dazzling displays of strength. But as ancient woodlands vanish, this forest gladiator faces decline.

Steneosaurus bollensis

This Jurassic marine crocodile hails from the Holzmaden deposit in Germany. Its exceptionally preserved fossils sometimes include skin and muscle, providing rare insight into marine life and predator adaptations about 180 million years ago. As a top predator of its time, *Steneosaurus* bollensis highlights the diversity and complexity of Jurassic marine ecosystems.

Tuatara

Sphenodon puncatus

The tuatara, a living fossil from the dinosaur age, stands alone in its ancient reptile lineage. Its dual upper teeth rows and sliding jaw create a powerful bite capable of shearing bone.

Panther chameleon
Furcifer pardalis

Panther chameleons are dazzling communicators, flashing bright colours to signal dominance or courtship and dark tones when stressed. More than camouflage, their shifting hues reveal a vivid language of mood and intent.

Eastern Crested Guineafowl

Guttera pucherani

The eastern crested guinea fowl, known for its curly black crest, roams in noisy flocks, using loud calls for defence and powerful legs to sprint through forests rather than fly.

Steneosaurus bollensis

This species of Jurassic crocodile was well-adapted to living in a marine environment, even more so than today's saltwater crocodiles. We find their fossils in deep ocean deposits, suggesting they swam far away from shore to hunt for prey.

Tropidaster sp. and *Palaeocoma* sp.

To the left is a large fossil of a regular starfish (*Tropidaster*) lying beside a group of brittle starfish (*Palaeocoma*). Found in a perfectly rounded Jurassic beach pebble from Dorset, UK, this fossil required many hours of skilled preparation to reveal its mesmerizing detail. The descendants of both species still thrive in today's oceans.

Socotra Cormorant

Phalacrocorax nigrogularis

The Socotra cormorant, found in the Arabian Gulf and Arabian Sea, dives over 18 metres to catch shoaling fish, resulting in it regularly getting caught in fishing nets. Conservation efforts with local fishermen are helping protect this remarkable marine bird.

Northwest Africa 12691 lunar meteorite

This rare 13.5-kilogram lunar meteorite from Western Sahara holds fragments of the Moon's surface, preserved in a breccia rock. Adhering orange soil marks its Earthly journey after thousands of years orbiting space.

Eagle Station pallasite meteorite

This rare Eagle Station pallasite, found in Kentucky in 1880, features large iron-rich olivine crystals embedded in nickel-iron metal, with remnants of orange soil showing its journey on Earth. It is the type specimen of a unique subgroup from a distinct asteroid, revealing extraordinary asteroid diversity.

Acutiramus macrophthalmus

One of the largest eurypterids, this impressive specimen grew to nearly 2 metres long and was found in the Silurian rocks of New York State. Composed of 23 segments, this giant predator likely drove early fish evolution, but all eurypterids went extinct at the end of the Permian Period, some 251.9 million years ago.

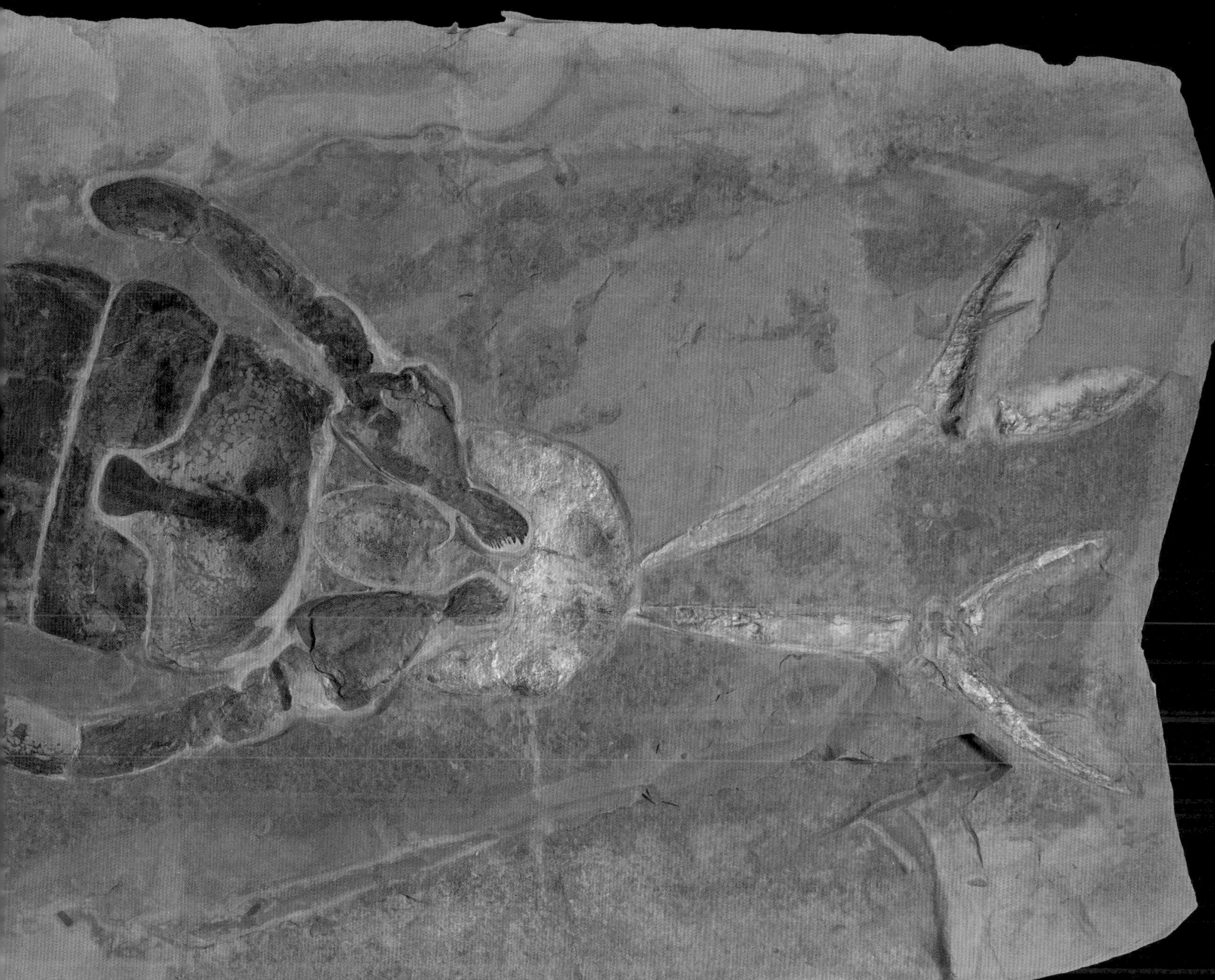

Grey-winged trumpeter

Psophia crepitans

In the dense Amazon rainforest, the grey-winged trumpeter acts as a vigilant sentinel, issuing sharp alarm calls to warn of approaching predators. Eating mainly fruit, they are vital seed dispersers. Their rare mating system features dominant females mating with multiple males, with group members helping raise chicks.

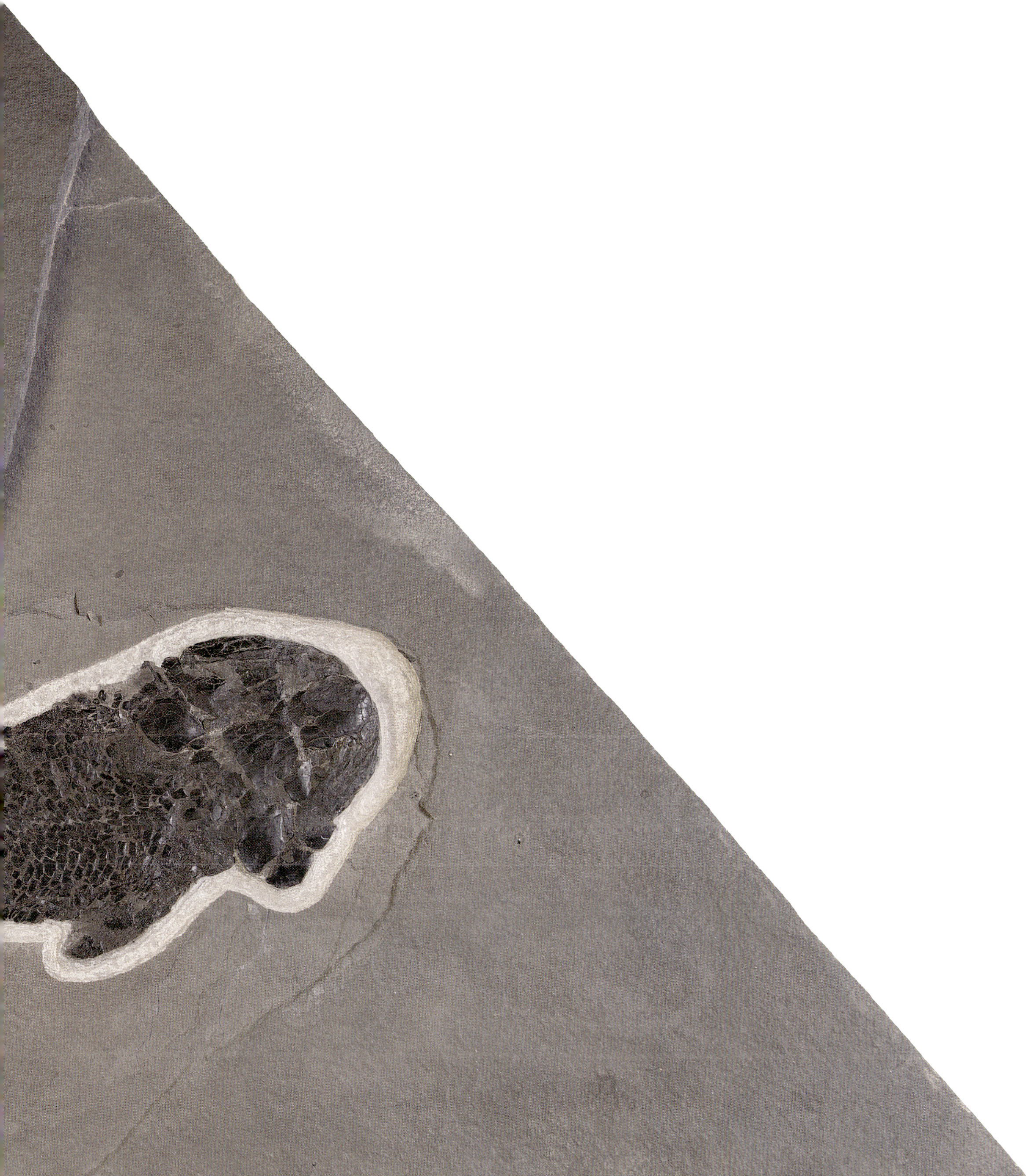

Gyroptychius agassiza

This fossil fish from the Devonian Period of Scotland and is approximately 380 million years old. This type of fish is closely related to us, as it belongs to a group of lobe-finned fish that gave rise to the first land animals.

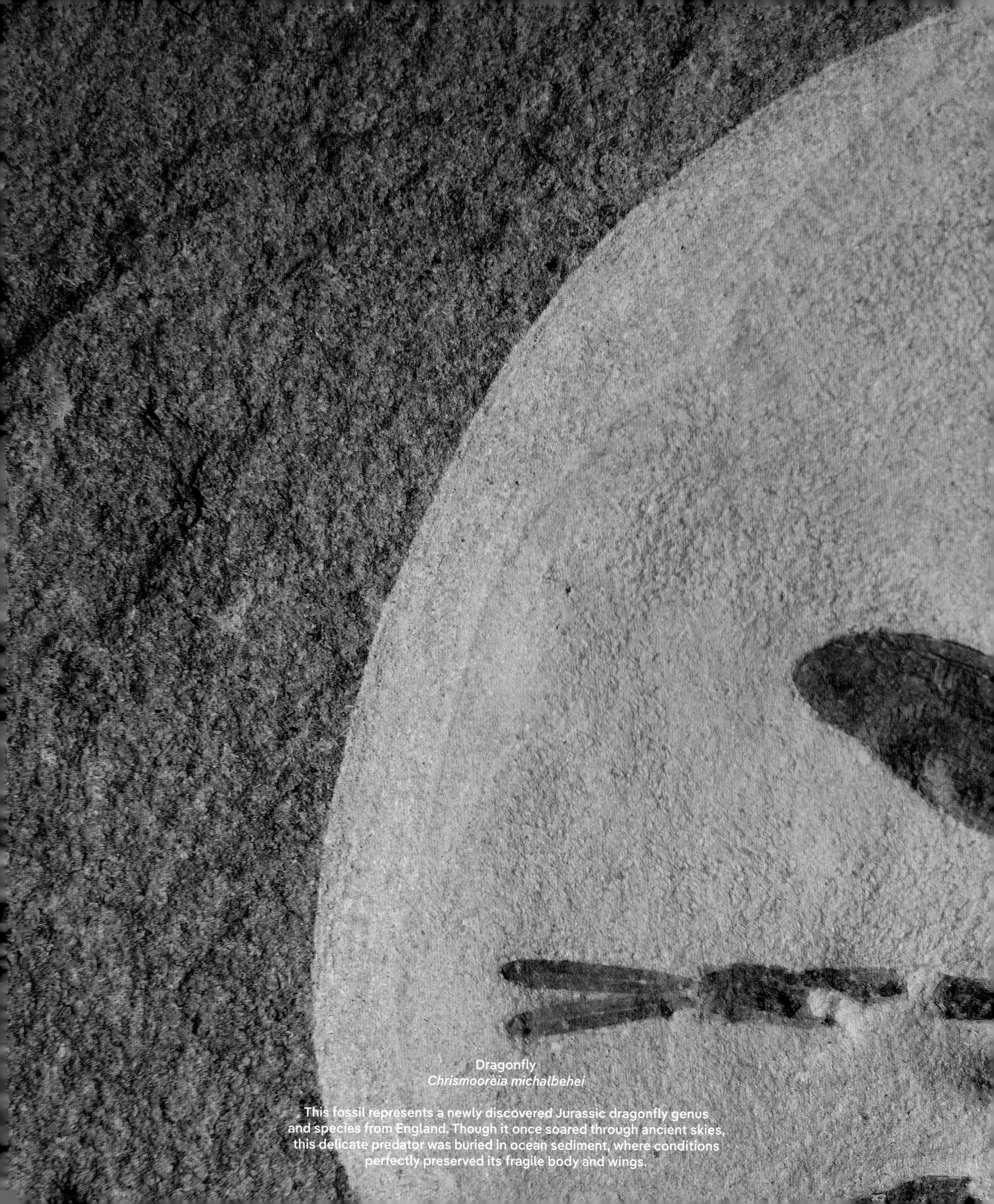

Dragonfly
Chrismooreia michalbehei

This fossil represents a newly discovered Jurassic dragonfly genus and species from England. Though it once soared through ancient skies, this delicate predator was buried in ocean sediment, where conditions perfectly preserved its fragile body and wings.

Black-throated loon
Gavia arctica

The black-throated loon, a skilled diver, breeds on deep freshwater lakes but winters along sheltered coasts, migrating as far south as the Mediterranean and Black Sea, embodying resilience in harsh northern habitats.

Arctic hare
Lepus arcticus

Admirably suited to the harsh Arctic tundra, the Arctic hare boasts dense seasonal fur, compact features to minimize heat loss, significant fat reserves, and large furry feet to navigate snow, while seeking shelter in dens for warmth.

Arctic fox
Vulpes lagopus

The specimen was modelled after the iconic image of an Arctic fox mid-leap—a dramatic hunting move known as 'moussing'. Adapted to extreme cold, it boasts dense fur, snowshoe-like paws, and seasonal colour changes, traveling vast distances to hunt or cache food.

Fossilised Ribcage and backbone of a Plesiosaur
Plesiosaurus sp.

This is the partial ribcage and backbone come from what was an articulated skeleton of long-necked, barrel-bodied predators of the Jurassic seas. The fossil may represent remains of a larger ichthyosaur's attack, showcasing the intense predator-prey dynamics of ancient seas.

Camarasaurus sp.

A Jurassic sauropod from southern Wyoming (approximately 150 million years old), bore a large skull and sturdy teeth, specialised for processing coarse vegetation unlike the slender-toothed Diplodocus.

Paracoroniceras lyra

Exemplifying Jurassic elegance, *Paracoroniceras lyra* is a 120-kilogram fossil block containing three exceptionally rare ammonites from the UK's Lower Lias, first unearthed in 2008 and once weighing over 200 kilograms.

Fossilised stingray
Cyclobatis major

This stingray-like fish from Lebanon's Haqil Formation, dating to roughly 95 million years ago, reveals exquisite preservation typical of the region's fossil reefs. Flattened in life and further compressed in stone, it endures like a delicate watercolour on rock.

Barbary falcon
Falco peregrinus pelegrinoides

The Barbary falcon, a desert-dwelling subspecies of peregrine falcon, inhabits rocky cliffs and mountains but also nests on urban towers. Highly prized in Middle Eastern falconry traditions, they're valued for their speed, stamina, and ability to hunt desert game birds.

Edmontosaurus sp. (right)

Exceptionally pale and resembling modern bone, this rare dinosaur fossil represents the most complete skeleton recovered from the Late Cretaceous Foxhills Formation—an invaluable specimen for museum study.

Following pages

Shoulder joint (top left)
The preserved shoulder bones of an *Edmontosaurus* uncovered in the Foxhills Formation, reveal an extraordinary creature from the Late Cretaceous Period. With a limb structure suggesting it primarily walked bipedally but could also move quadrupedally, *Edmontosaurus*'s specialised anatomy highlights its adaptation to a dynamic lifestyle millions of years ago.

White hadrosaur jaw (top right)
Part of a vast group of *hadrosaurs*, the left lower jaw (dentary) of this *Edmontosaurus* reveals the hundreds of tightly packed teeth it used to grind the tough vegetation it fed on during the Late Cretaceous Period.

Spinal bone (bottom left)
A single tail bone (vertebra) reveals a tall bony spine that would have been covered in muscle and skin, creating a long ridge running down the tail of this hadrosaur dinosaur.

Tibia (bottom right)
The tibia, the larger of the two lower leg bones from the white hadrosaur skeleton found in South Dakota, reveals a robust, sturdy build in this herbivorous dinosaur—strong but not designed for speed or agility.

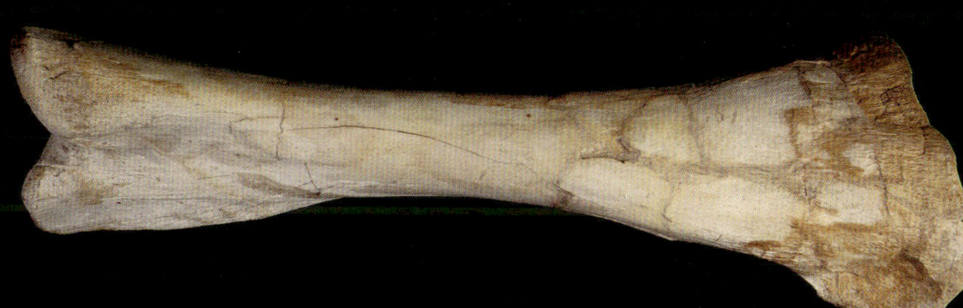

Greater celandine

Chelidonium majus

Greater celandine was among the first plants formally named by Carl Linnaeus in 1753. Native to much of Europe, it also grows widely throughout the Middle East, reflecting a long history of botanical recognition and a broad natural distribution.

Wooly burdock

Arctium tomentosum

The woolly burdock is named for its soft, wool-like leaves and cobweb-like hairs on the flower heads, but caution is advised when close to its burrs, which can easily cling to clothing and skin.

Purple-winged grasshopper
Titanacris albipes

Most grasshoppers blend into foliage, but the purple-winged grasshopper dazzles with vivid underwings. The giant *Titanacris albipes*, spanning 200 mm, inhabits the vibrant Amazon rainforest and nearby regions.

Graceful decorator crab
Oregonia gracilis

Graceful yet resilient, the decorator crab embellishes its shell with algae and sponges using hooked bristles. These adornments offer camouflage and chemical defence. Dwelling in cold Pacific reefs and kelp forests, it withstands strong currents with its long, delicate legs.

Hyacinth macaw

Anodorhynchus hyacinthinus

The hyacinth macaw, the world's largest flying parrot, possesses one of the strongest beaks in the avian realm—powerful enough to crack Brazil and palm nuts. By dispersing partially digested seeds, it helps regenerate tropical woodlands.

Superb starling

Lamprotornis superbus

Radiant and social, the superb starling owes its iridescent plumage to melanin-based nanostructures that refract light, not pigment. Its orange belly derives from its diet, and its lively songs and mimicry enrich its social life.

Madagascan sunset moth
Chrysiridia rhipheus

Though it resembles a butterfly, this is a day-flying moth. Its dazzling iridescence comes from light scattering through microscopic wing scales. In Madagascar, vast swarms migrate to locate host plants for their larvae.

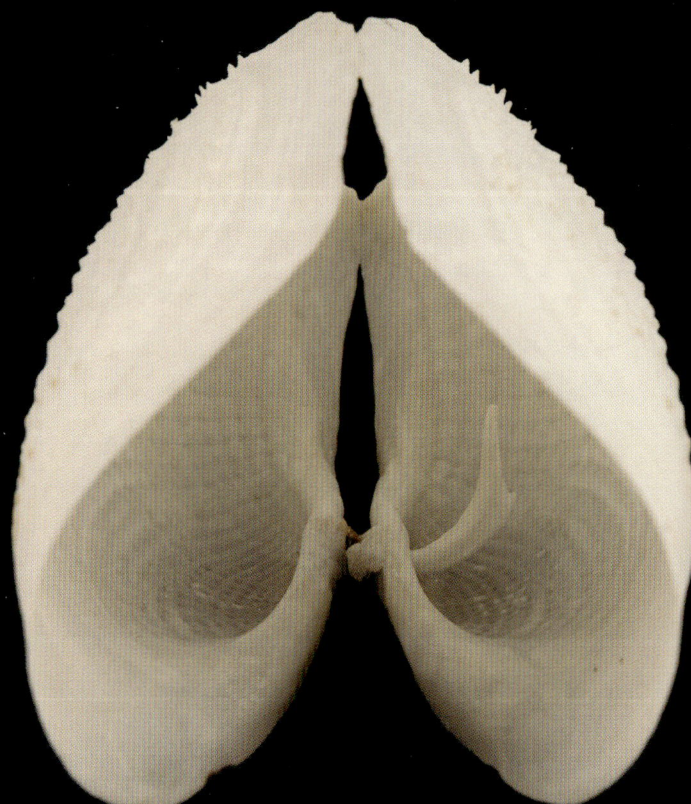

'Angel wing' shell
Barnea dilatata

Living buried in sand up to 20 metres deep, this striking white bivalve filter-feeds through an exceptionally long siphon prized as a delicacy in regions like Japan's Ariake Sea. Its delicate, wing-like shell—often reaching over 120 mm—makes it a collector's favourite.

Sea beans
Mucuna elliptica

Sea beans are drifting seeds from a 30-metre tropical vine with bristly stems, lilac-white blooms and dark pods, carried by ocean currents from Africa and Asia to distant islands.

Demoiselle crane

Anthropoides virgo

Small in stature but mighty in endurance, the demoiselle crane crosses the Himalayas each year at heights exceeding 4,500 metres. Migrating in synchronised flocks, it conserves energy and express strong pair bonds through graceful courtship dances of leaps, bows, and airborne grass.

Petrified wood

All life on Earth has the possibility of becoming a fossil, and this includes plants. The fibrous and often tough remains of trees make it into the fossil record, as did this 200+ million-year-old log from the Triassic Period of Arizona in the USA.

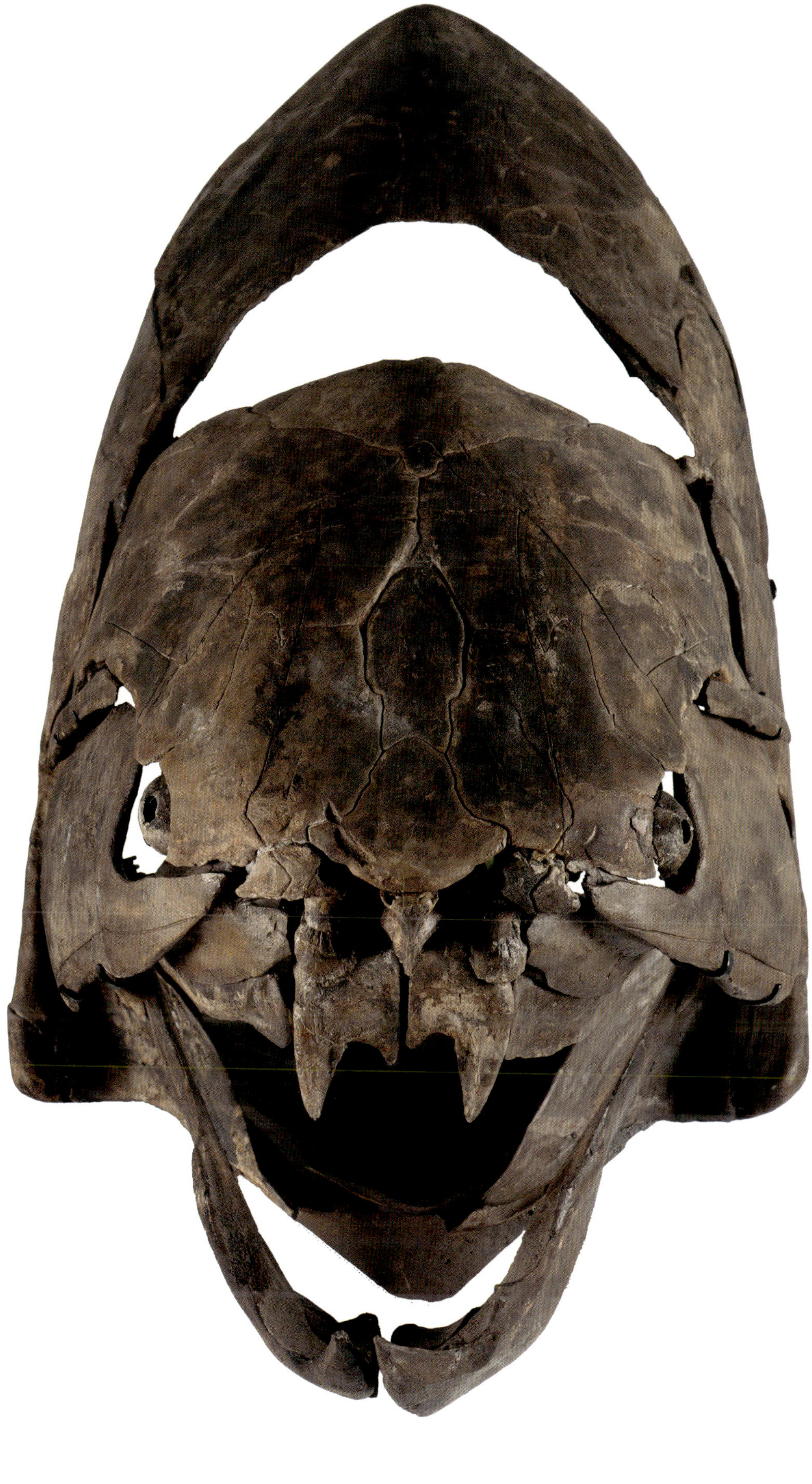

Dunkleosteus sp.

Among the earliest ocean apex predators, these heavily armoured fish from 380 million years ago evolved against giant sea scorpions, using bony plates as razor-sharp jaws to slice prey.

Giant isopod
Bathynomus giganteus

Among the largest isopods at up to 50 centimetres, this species thrives in deep-sea extremes, slowly scavenging on sunken organic matter and surviving months without food due to its low metabolism.

Asiatic lion
Panthera leo persica

The Asiatic lion, the only lion subspecies beyond Africa, now survives solely in Gujarat's Gir Forest in India. About 600 remain, marked by a distinctive belly fold and shorter, ear-revealing mane.

Greater argonaut

Argonauta argo

Found in warm seas worldwide, the greater argonaut is a pelagic octopus whose females reach 30 centimetres and craft paper-thin shells closely resemble a nautilus shell to project their floating eggs, unlike the tiny, shell-less males.

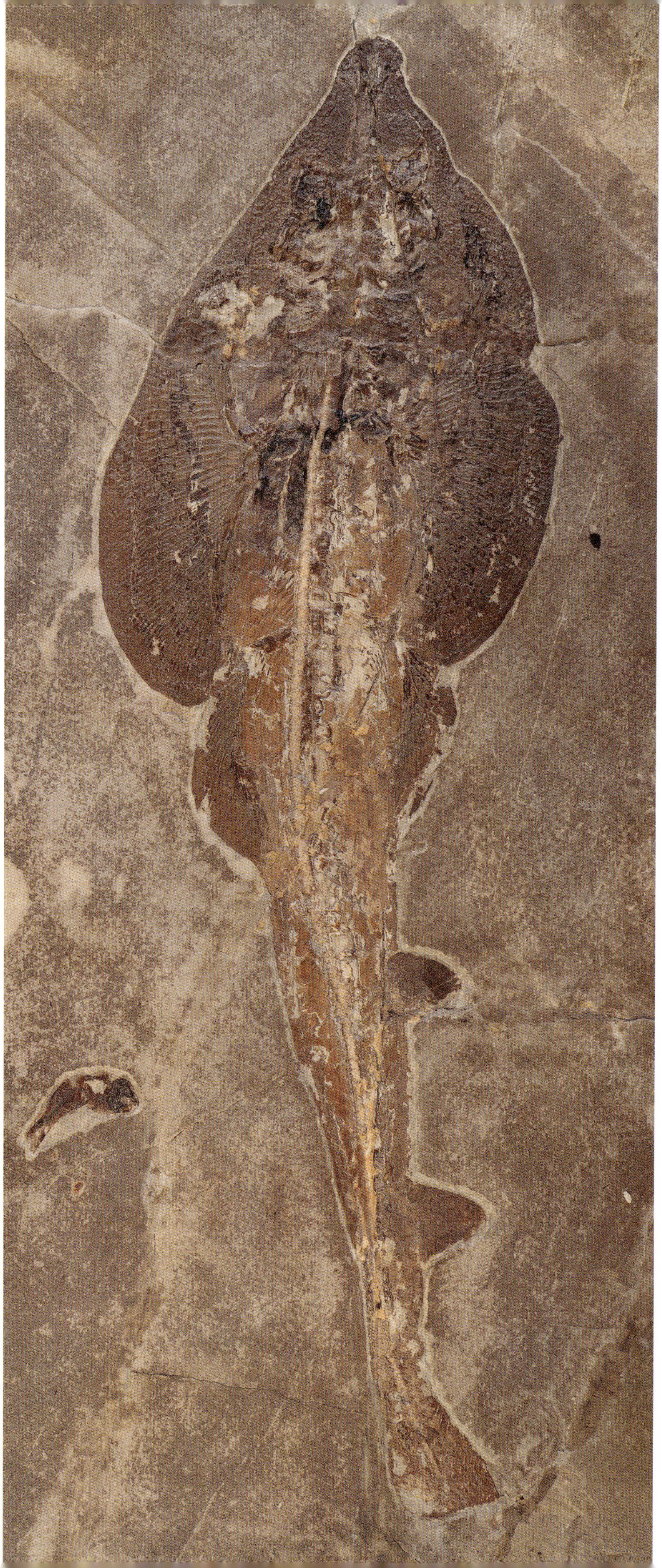

Guitar fish

Rhombopterygia rajoides

The preservation of the Cretaceous-aged fossils of Lebanon is very distinctive, often preserving soft tissues, such as skin and muscle, that can help provide a better picture of what these animals might have looked like in life. The amazing fossils of the guitar fish are even more remarkable, as these belonged to the boneless shark family whose skeletons were made of soft cartilage and should not have made it into the fossil record.

Saurorhynchus acutus

The seas of Britain during the Lower Jurassic teemed with fierce predators and the creatures they hunted. This skull belonged to a fish that likely became a meal for the mighty ichthyosaurs and plesiosaurs—the ancient sea dragons of a vanished world.

Fossilised plant

Otozamites sp.

Beautifully preserved plants drifted out on gentle currents but then sank to the sea floor to be preserved as fossils. This delicate fossil tells us that cycads were growing on land, while sea dragons ruled the oceans.

Stegosaur

The armour-plated dinosaurs of the Jurassic Period included *Stegosaurus* and *Hesperosaurus*, but this stegosaur skeleton looks different from these known species and looks to be new to science. Once the skeleton has been fully researched and described by scientists, we can give this a new name. Its blood-vessel-rich plates likely aided temperature control, while its spiked tail offered defence.

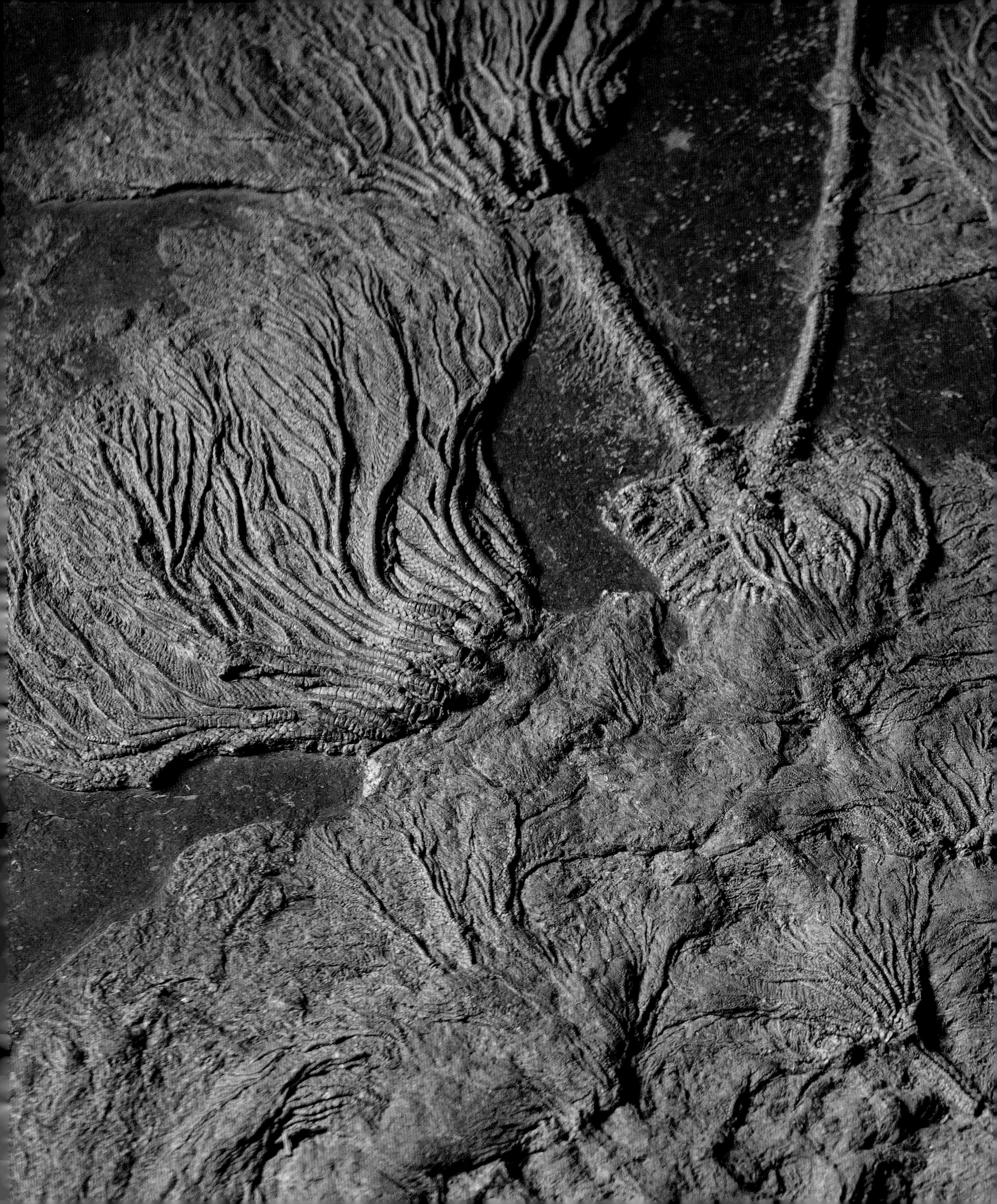

Crinoid slab
Seirocrinus subangularis

Colonies of this Jurassic crinoid, a starfish relative, attached themselves to floating driftwood and were gently transported across oceans but some of the colonies ended up sinking to the seafloor to be preserved as these beautiful records of past life. While these animals are often called 'sea lilies' the only plant associated with this organism is the driftwood on which they relied upon for a home.

Dar al Gani 406 iron meteorite

The 5.1-kilogram flight-oriented iron meteorite was discovered in Libya's Al Jufrah region in 1998. One side has a rugged shrapnel structure, suggesting it was initially part of a much larger mass before breaking away during its fiery descent. The other side shows numerous shallow depressions and a typical brownish patina.

Dar al Gani 190 carbonaceous chondrite

This 2.58kilogram main mass of Dar al Gani 190, a rare CO3 carbonaceous chondrite found in Libya in 1996, holds ancient silicate chondrules that formed 4.5 billion years ago in the early Solar System.

Ammonite

Oxynoticeras lymense

The ammonites of Lyme Regis are renowned for their elegance. In this specimen, golden pyrite outlines each chamber, while translucent calcite fills the spaces, allowing light to pass through these stunning fossils when they are polished like this one.

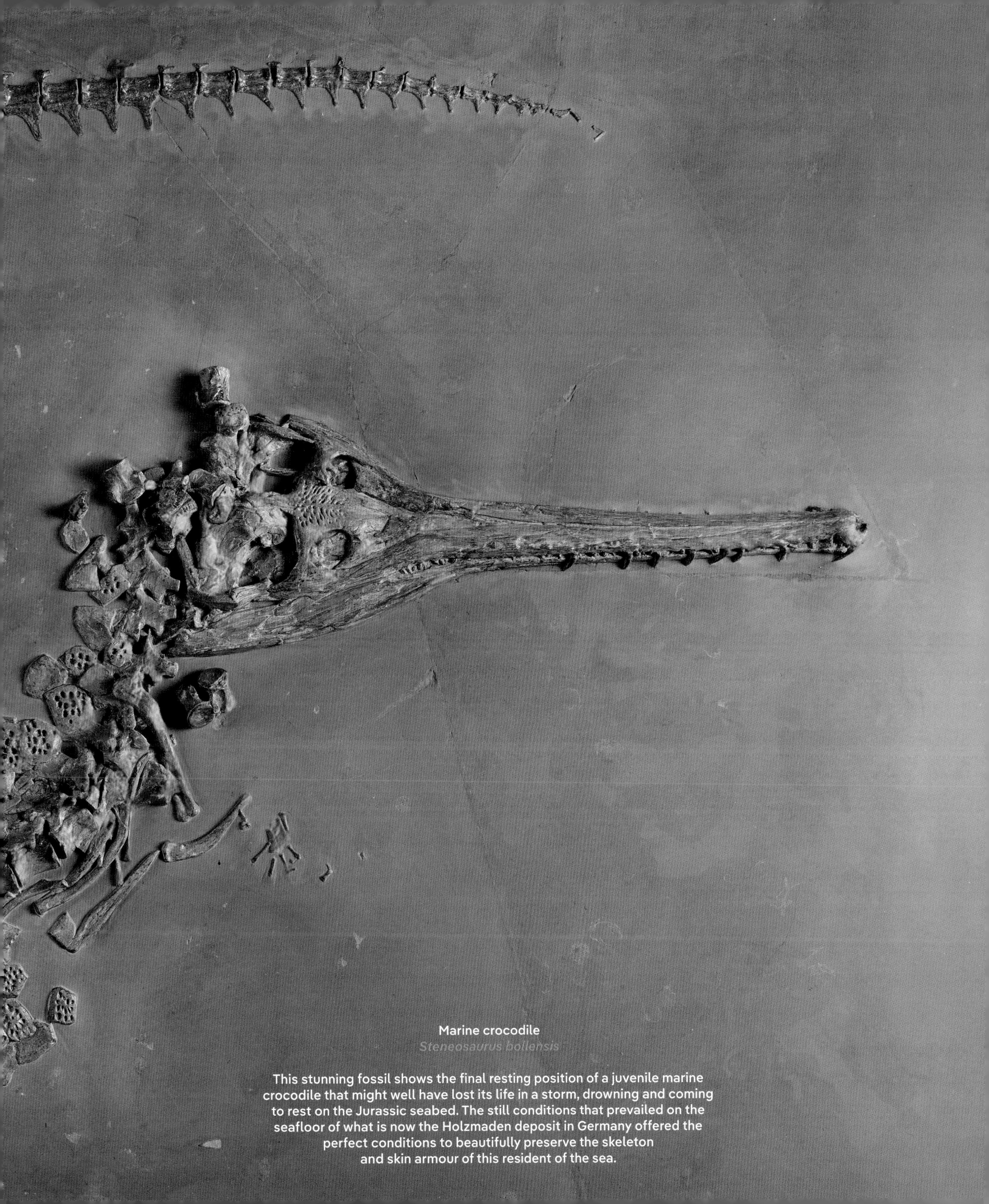

Marine crocodile

Steneosaurus bollensis

This stunning fossil shows the final resting position of a juvenile marine crocodile that might well have lost its life in a storm, drowning and coming to rest on the Jurassic seabed. The still conditions that prevailed on the seafloor of what is now the Holzmaden deposit in Germany offered the perfect conditions to beautifully preserve the skeleton and skin armour of this resident of the sea.

Emperor scorpion (female)
Pandinus imperator

Native to West and Central Africa, the emperor scorpion ranks among the world's largest. Young scorpions rely on their sting to subdue prey, while adults depend on their powerful pincers.

Trilobites
Dikelokephalina brenchleyi and *Asaphellus stubbi*

This remarkable slab holds a dense cluster of trilobites, a 480-million-year-old gathering frozen in time. Discovered in Ouled Slimane, Morocco, it preserves over forty fossils from two coexisting species.

North Island brown kiwi
Apteryx mantelli

The North Island brown kiwi, cloaked in soft, hair-like feathers, prowls night forests on sturdy legs, nose to earth, seeking life below, and tending its single, enormous egg—a quiet marvel of devotion. Kiwi eggs are enormous compared to the bird's body size—up to 20% of the female's body weight.

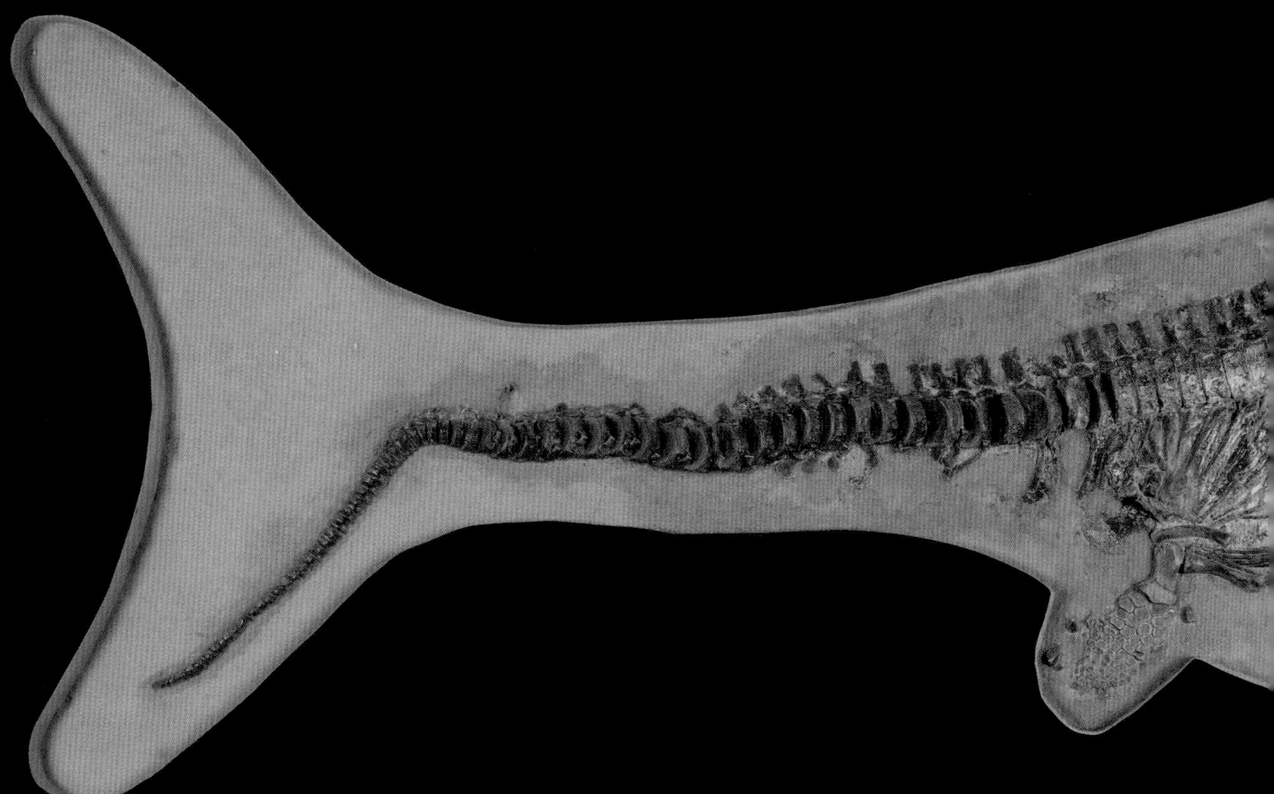

Ichthyosaurus communis

Ichthyosaurus, meaning 'fish-lizard', tells the story of land creatures that returned to the sea to feast on its rich bounty. The large circle of bones on each side of the skull would have held an enormous eye indicating this predator had acute vision, making it the perfect undersea hunter.

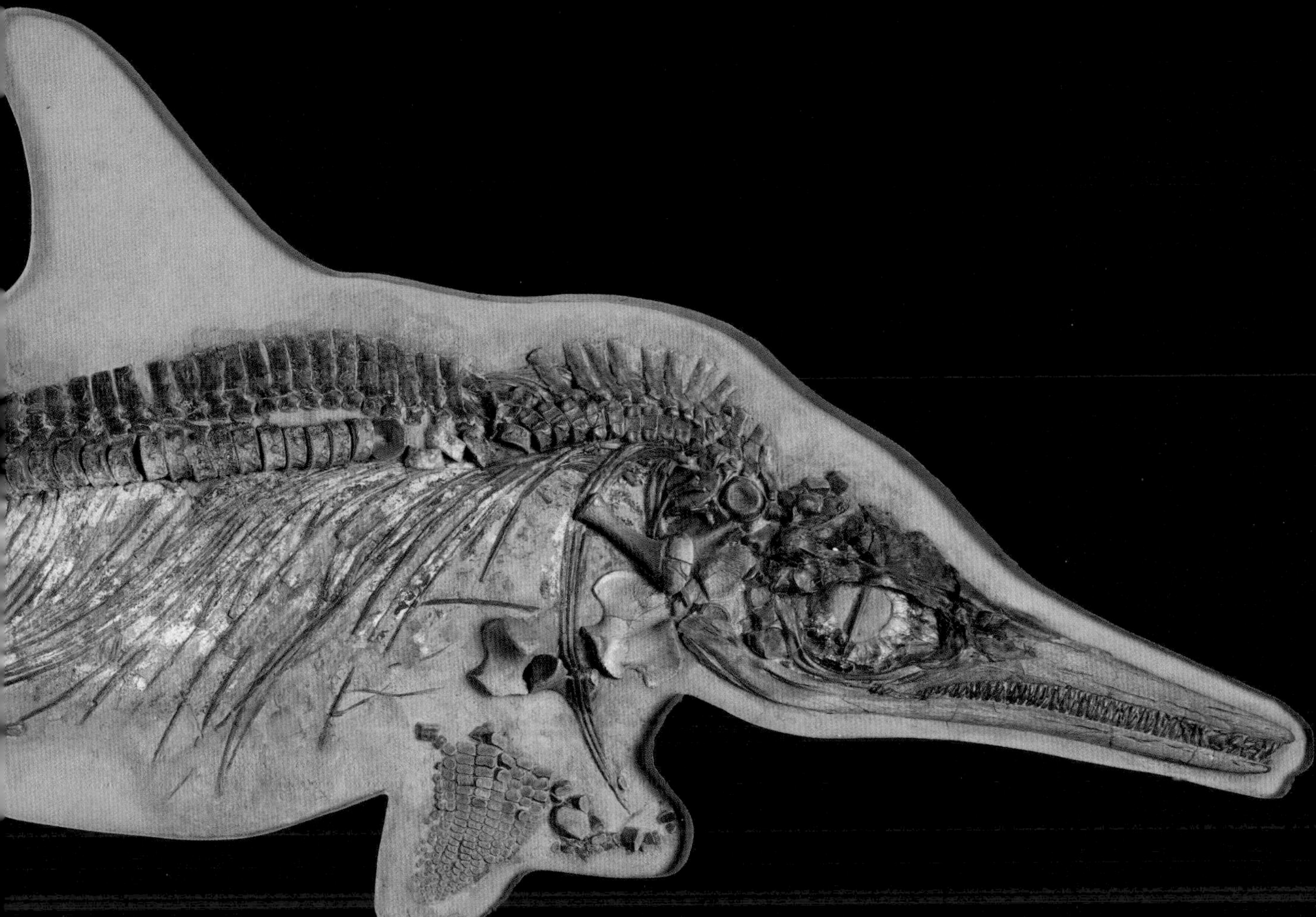

Fern

Remia pinnatifida

This ancient fern dominated Earth's landscapes some 310 to 280 million years ago, leaving its fossilised traces scattered across Europe. It marks a fascinating turning point in Earth's history—a time when lush, swampy forests began giving way to drier, more temperate environments.

Annularia stellata

Annularia, an extinct carboniferous plant, displayed elegant whorls of 8–13 flat leaves encircling tall calamites—towering stems that once thrived in ancient swamp forests 300 million years ago.

Sea dragon and turtle fossil

Tylosaurus kansasensis and *Chelophargis advena*

This striking fossil duo captures a timeless scene—predator and prey locked in the immortal battle of the hunter and the hunted. The marine reptiles of the Kansas Chalk from 85 million years ago are typified by the beautifully preserved skeleton of *Tylosaurus*. It is no surprise that these mighty predators are often called sea dragons. The turtle (*Chelosphargis*) that is swimming from the jaws of *Tylosaurus*, provides us with a rare glimpse of one of the most complete skeletons known of this extinct sea turtle.

Curl-crested aracari

Pteroglossus beauharnaisii

Distinctive for its curled head feathers, the curl-crested aracari uses these ornamentals in courtship and social signalling. By feasting on fruit and dispersing seeds, it sustains forest renewal.

Tiger pistol shrimp

Alpheus bellulus

The tiger pistol shrimp wields a remarkable snapping claw that fires with explosive force, creating shockwaves and flashes of light capable of stunning prey—its underwater snap exceeding 200 decibels.

Purple Spot Mantis Shrimp
Gonodactylus smithii

The mantis shrimp wields lightning-fast raptorial limbs, striking at 23 metres per second with bullet-like acceleration. Each hit issues collapsing cavitation bubbles that unleash shockwaves and sun-hot flashes, stunning prey. With vision across 12 colour channels, it masters the kaleidoscopic world of coral reefs.

Cushion starfish and Brittle star fish
Ophioderma

The curled movement in the arms of the Jurassic cushion starfish was frozen in time as this community was buried in sediment some 190 million years ago. The whole slab preserves a snapshot in time when surrounding brittle starfish (*Ophioderma*) were also immortalised in stone. It is rare that the preservation, anatomy and positioning of fossils bring them so vividly back to life.

Harpy eagle

Harpia harpyja

The harpy eagle ranks among the largest and heaviest raptors globally, with a wingspan of up to 201 centimetres and a body weight of about 6 to 9 kilograms in the significantly larger females. A true apex predator, it hunts monkeys, sloths, and armadillos. Early Spanish explorers in Venezuela likened its majesty to the mythical griffon—half lion, half eagle.

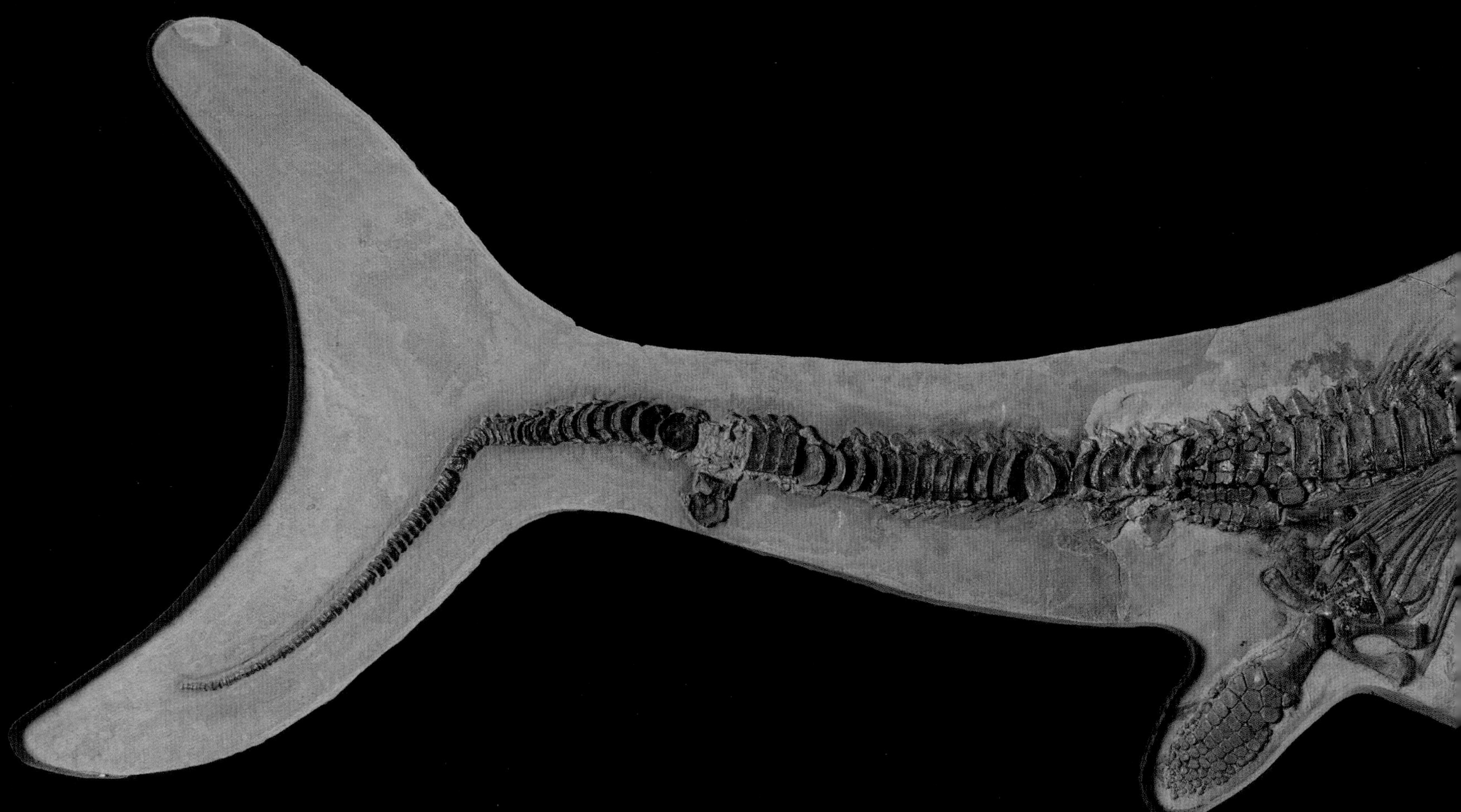

Marine reptile
Ichthyosaurus somersetensis

This Jurassic Sea dragon is from the coastal town of Watchet in Somerset (UK). The fossil preserves its final meal of squid and cuttlefish, complete with an inky stain preserved in its gut, a rare snapshot of ancient marine life, frozen in time.

Following pages

Venus sea fan
Gorgonia flabellum

Commonly called the Venus sea fan, this sessile soft coral forms broad, fan-shaped colonies of tiny polyps set within a flexible skeleton of calcite and collagen. Each eight-tentacled polyp filters plankton, with colonies growing perpendicular to currents to feed most efficiently.

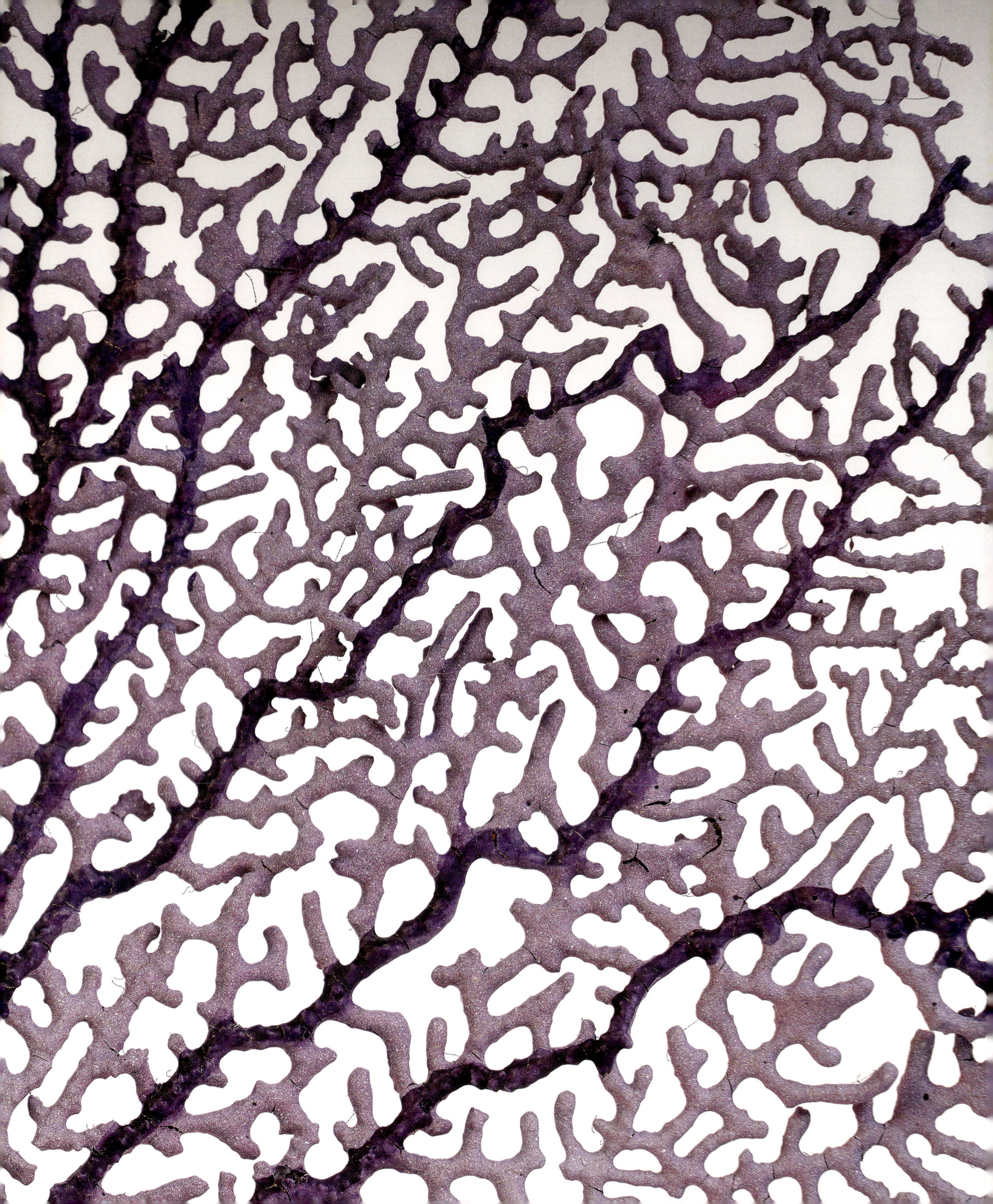

Caracal

Caracal caracal

Renowned for leaping up to 3 metres to catch birds mid-flight, caracals are incredible jumpers. Historically, they were trained by royalty in India and Persia for bird hunting and held cultural importance in ancient Egypt as guardians.

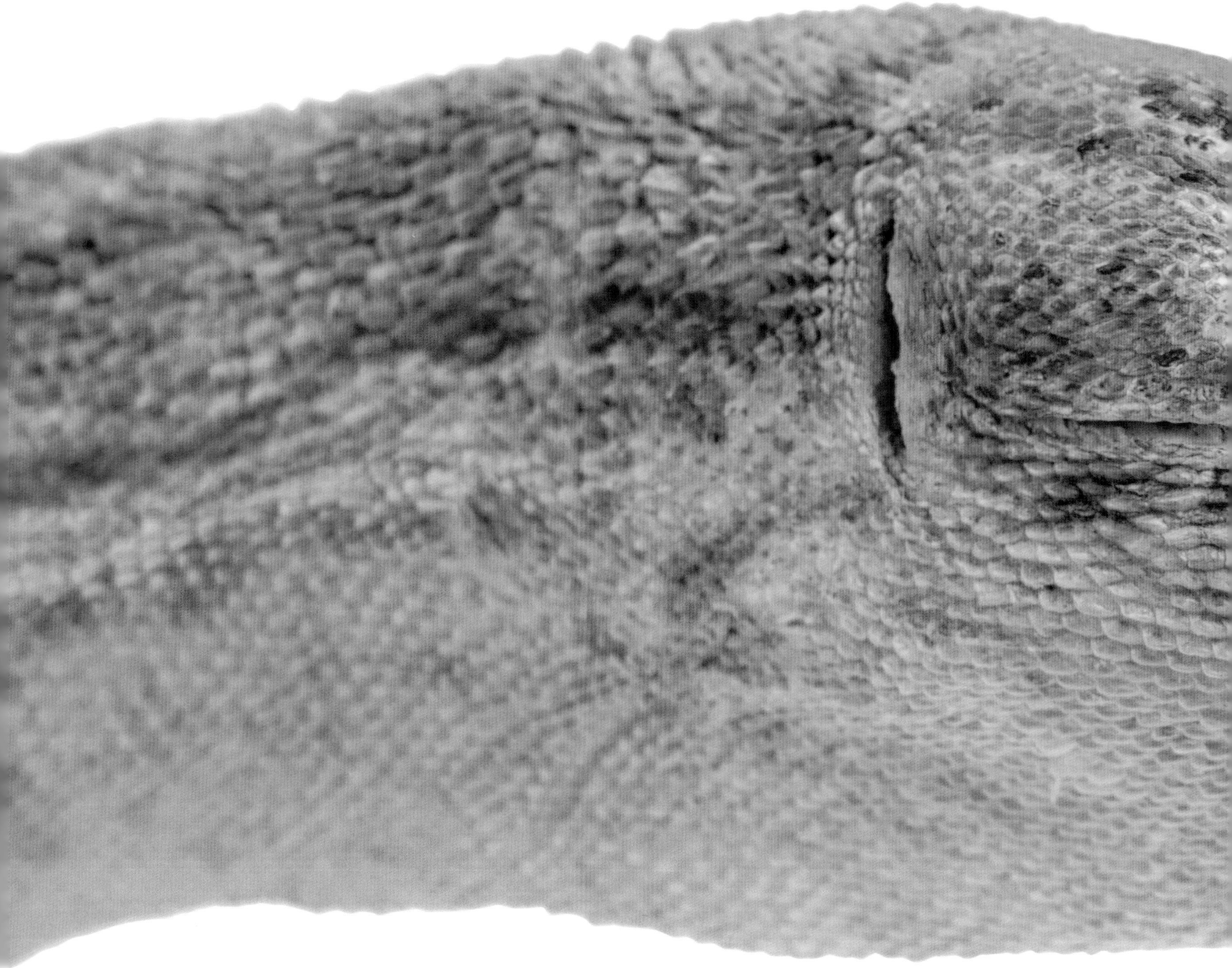

Desert Monitor

Varanus griseus

The desert monitor thrives in harsh deserts by regulating its temperature behaviourally. It keeps its body near 41°C through hyperventilation and seeks shade or burrows, adapting well to extreme heat.

Eurypterus remipes

Eurypterus remipes belongs to a large and successful group of extinct marine arthropods. The name of this group (*Eurypterida*) comes from Ancient Greek meaning ‘broad’ or ‘wide’ and ‘wing’, referring to the pair of swimming paddles found in many species of marine organism. Fossils often show groups together, likely due to moulting gatherings washed into seafloor depressions, preserving moments of ancient life.

Huge Crinoid Fossil
Seirocrinus subangularis

Crinoids, ancient marine animals dating back over 500 million years, grew up to 4-metres long in Jurassic Holzmaden, Germany. They formed vast, intertwined colonies attached to driftwood, numbering in the hundreds, revealing vibrant prehistoric seascapes.

Krasnojarsk pallasite meteorite

This extraordinary 329-gram specimen formed at the boundary of an asteroid's metallic core and silicate mantle. Discovered as a 700-kilogram iron mass in Siberia in 1749, the famed 'Pallas iron' sparked lively scientific debate about native iron's origins. Named after the German naturalist Peter Simon Pallas, who championed its study, this meteorite lent its name to the pallasite group, forever marking a milestone in meteoritics.

Murchison carbonaceous chondrite

This 1.3-kilogram specimen fell in 1969 near the town of Murchison in south-east Australia. It contains 7-billion-year-old presolar grains and was the first to reveal extraterrestrial amino acids, nucleobases, and sugars—key life ingredients—offering crucial clues to life's origins on Earth and cosmic chemistry.

Ammonites
Asteroceras Obtusum

This exceptionally large specimen of *Asteroceras*, beautifully preserved in Jurassic rocks from Charmouth, Dorset, UK, was found loose on the beach—remarkably surviving decades of picnics and fossil hunting by visitors and tourists alike.

Grey crowned crane
Balearica regulorum

The grey crowned crane, crowned with a regal halo of golden feathers, captivates with graceful dances and playful social displays, standing elegantly as a symbol of Africa's vibrant wildlife.

Northwest Africa 10519 enstatite achondrite

This 533-gram specimen is a rare enstatite achondrite meteorite from Mauritania, shocked on its parent asteroid before ejection. As it entered Earth's atmosphere, its surface partially melted and ablated, creating characteristic dimples and depressions from turbulent airflow.

Blue whale
Balaenoptera musculus

Gentle giants of the ocean, blue whales play a vital role in marine life by circulating nutrients through their immense feeding and diving cycles, helping sustain plankton that fuel entire ecosystems. Without the blue whale to control their population, krill could consume so much phytoplankton that oxygen levels in the ocean could become unbalanced, adversely affecting marine life. Their graceful, powerful migrations span thousands of kilometres, and their haunting songs echo through the deep, symbolizing the ocean's vast, living heartbeat.

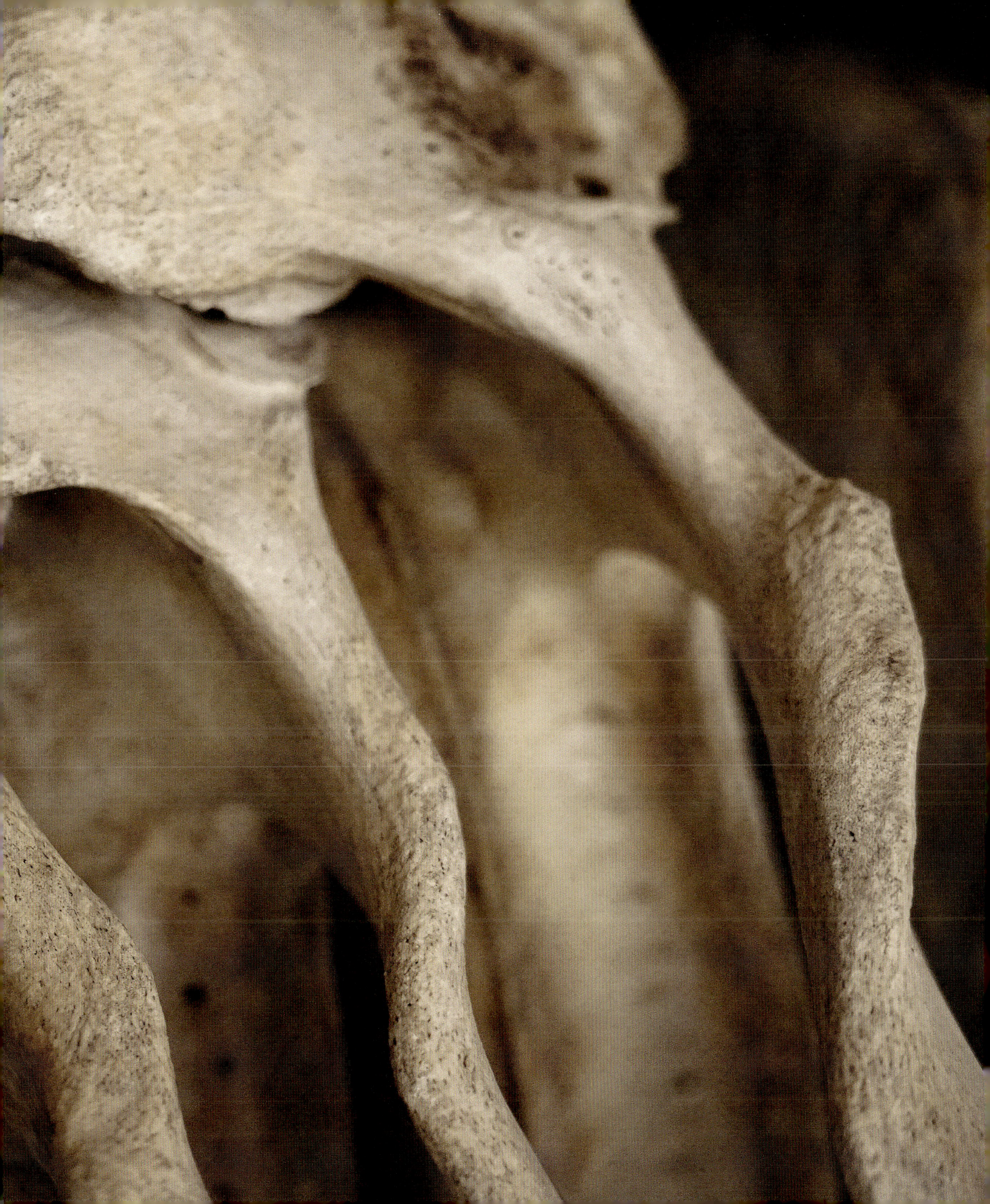

Julodis euphratica
Sulfurous jewel beetle

The Sulphurous jewel beetle, named for its shimmering wings
and bright yellow body, buzzes near plants in UAE and Oman wadis,
often first noticed by its deep humming flight sound.

Blotched blue-tongued lizard

Tiliqua nigrolutea

When threatened, the blotched blue-tongued lizard turns boldly, revealing its vivid blue tongue against a pink maw—a striking bluff that startles predators and masks its harmless nature with elegant defiance.

Brown-necked raven
Corvus ruficollis

Brown-necked ravens, desert specialists, display remarkable intelligence by cooperatively hunting lizards in pairs, blocking escape routes and striking as a united team—an elegant testament to survival in harsh landscapes.

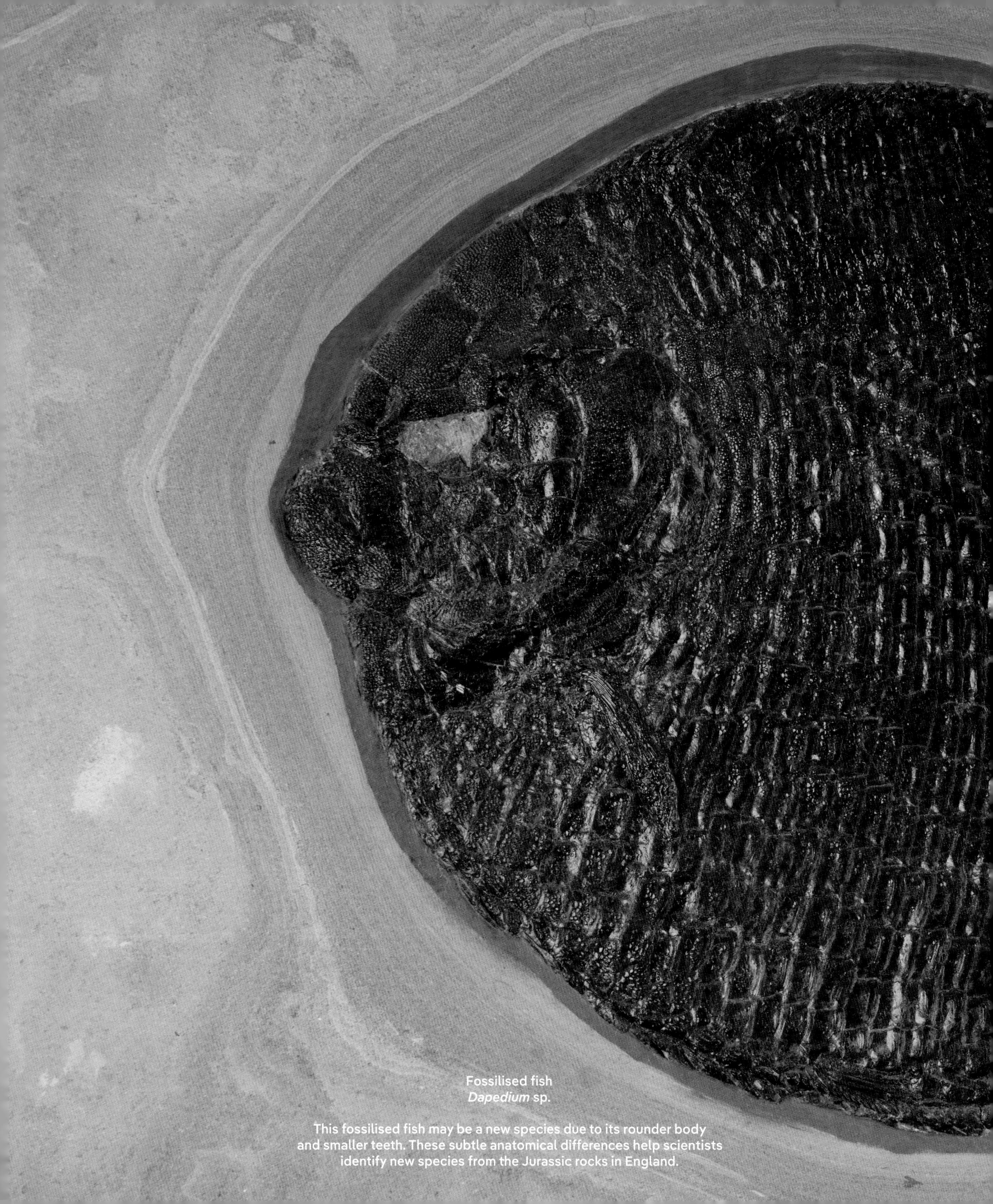

Fossilised fish
Dapedium sp.

This fossilised fish may be a new species due to its rounder body and smaller teeth. These subtle anatomical differences help scientists identify new species from the Jurassic rocks in England.

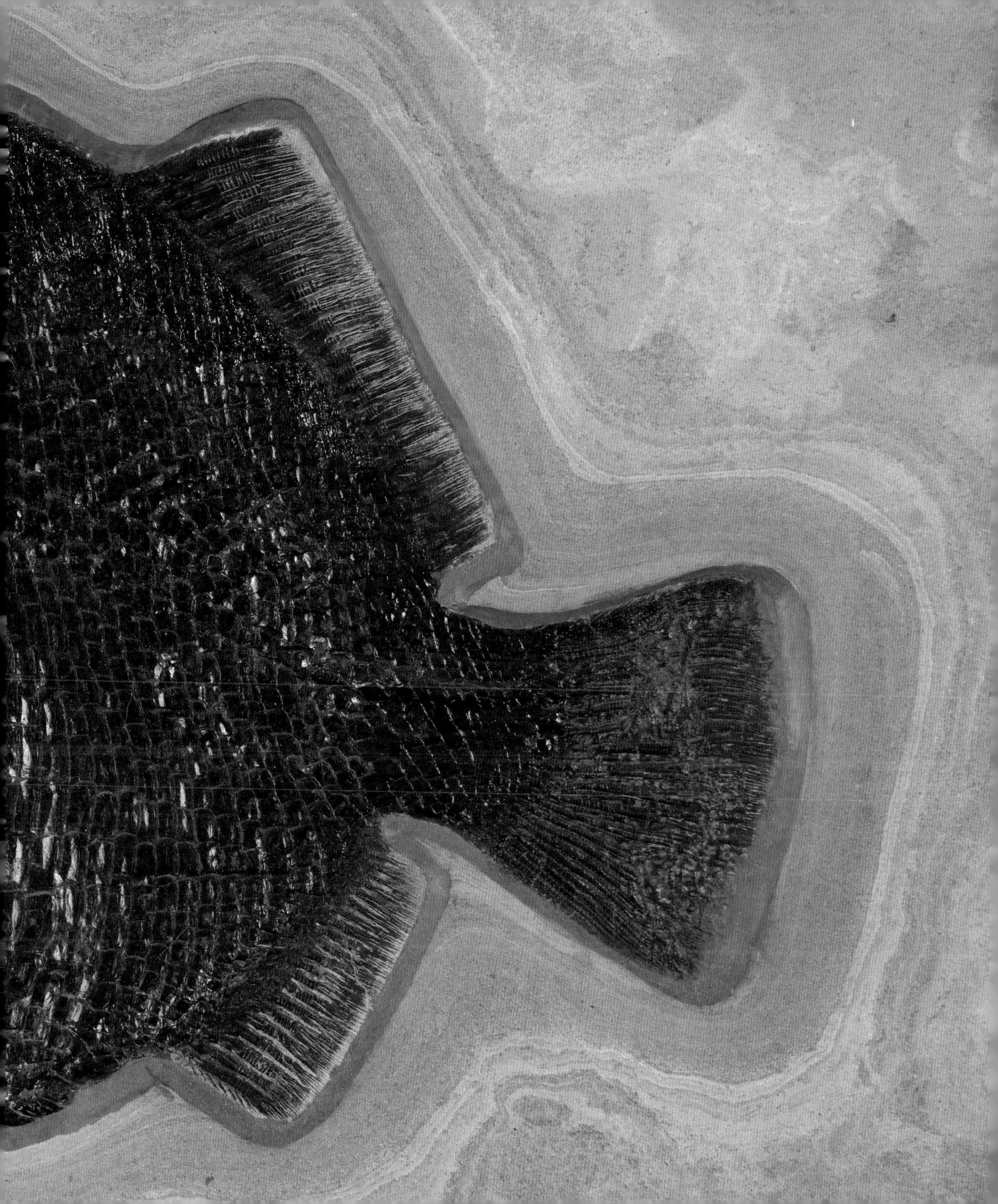

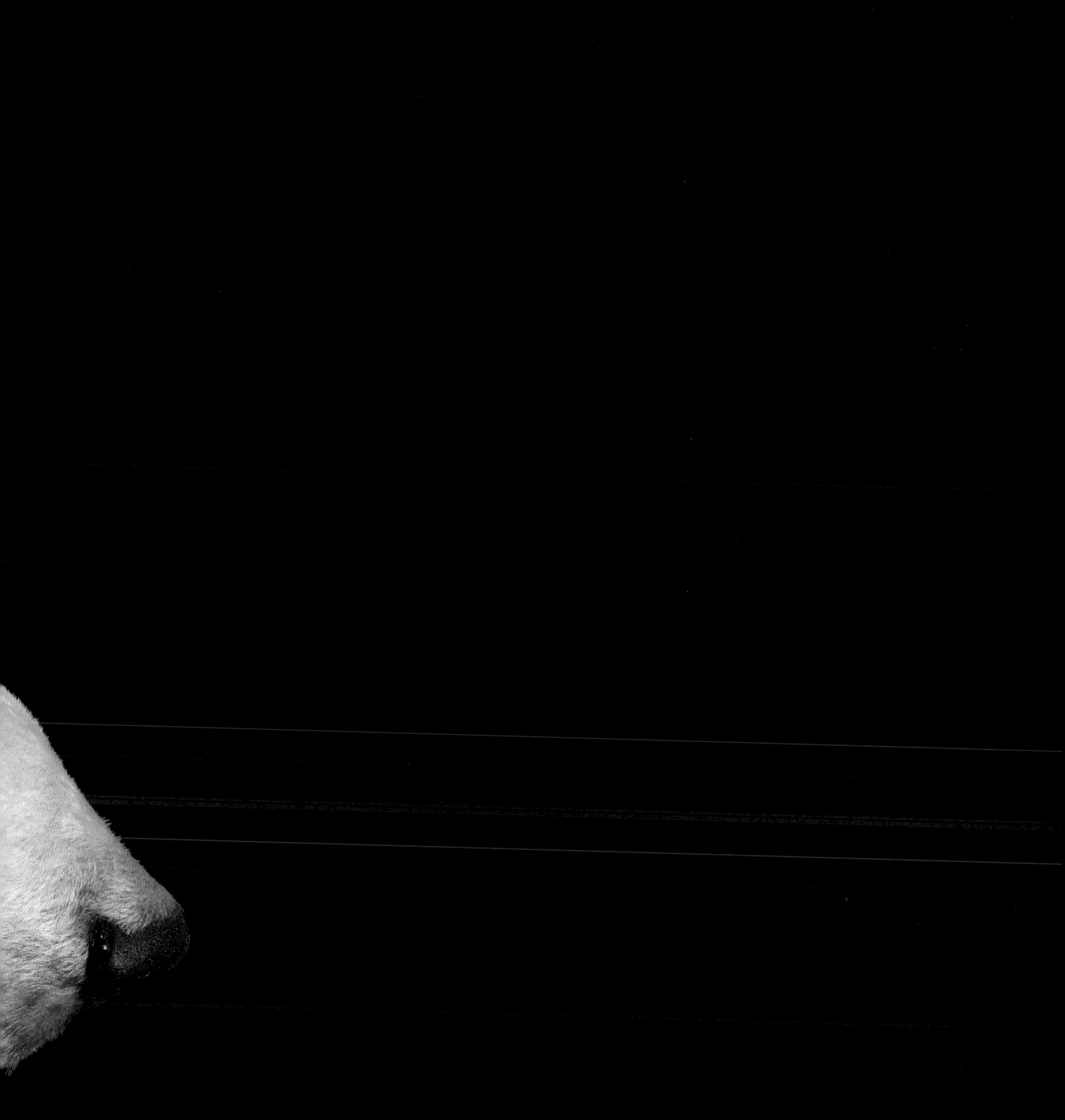

Polar bear
Ursus maritimus

Polar bears' transparent fur reflects light, while their black skin absorbs heat. Their powerful smell detects seals beneath ice and their large, webbed paws enable long, efficient swims in the Arctic's icy waters.

Tundra wolf

Canis lupus arctos

Arctic wolves have shorter muzzles, smaller ears and double-layered fur to minimize heat loss. They thrive in temperatures as low as -30°C. They live in packs of 4–7 members, with strong social bonds and cooperative hunting strategies. Their howls can travel up to 10 kilometres, helping coordinate hunts and maintain territory.

Diplodocid sauropod dinosaur

Diplodocid sauropods had long, whip-like tails, and at least fifteen neck vertebrae, when compared to mammals that only have seven. The late Jurassic fossil was excavated from a quarry in northern Wyoming, near the town of Tensleep, from the rocks of the Morrison Formation. The relatively small skull is lined with pencil-like teeth, only enabling this dinosaur to strip leaves off branches with its delicate jaws.

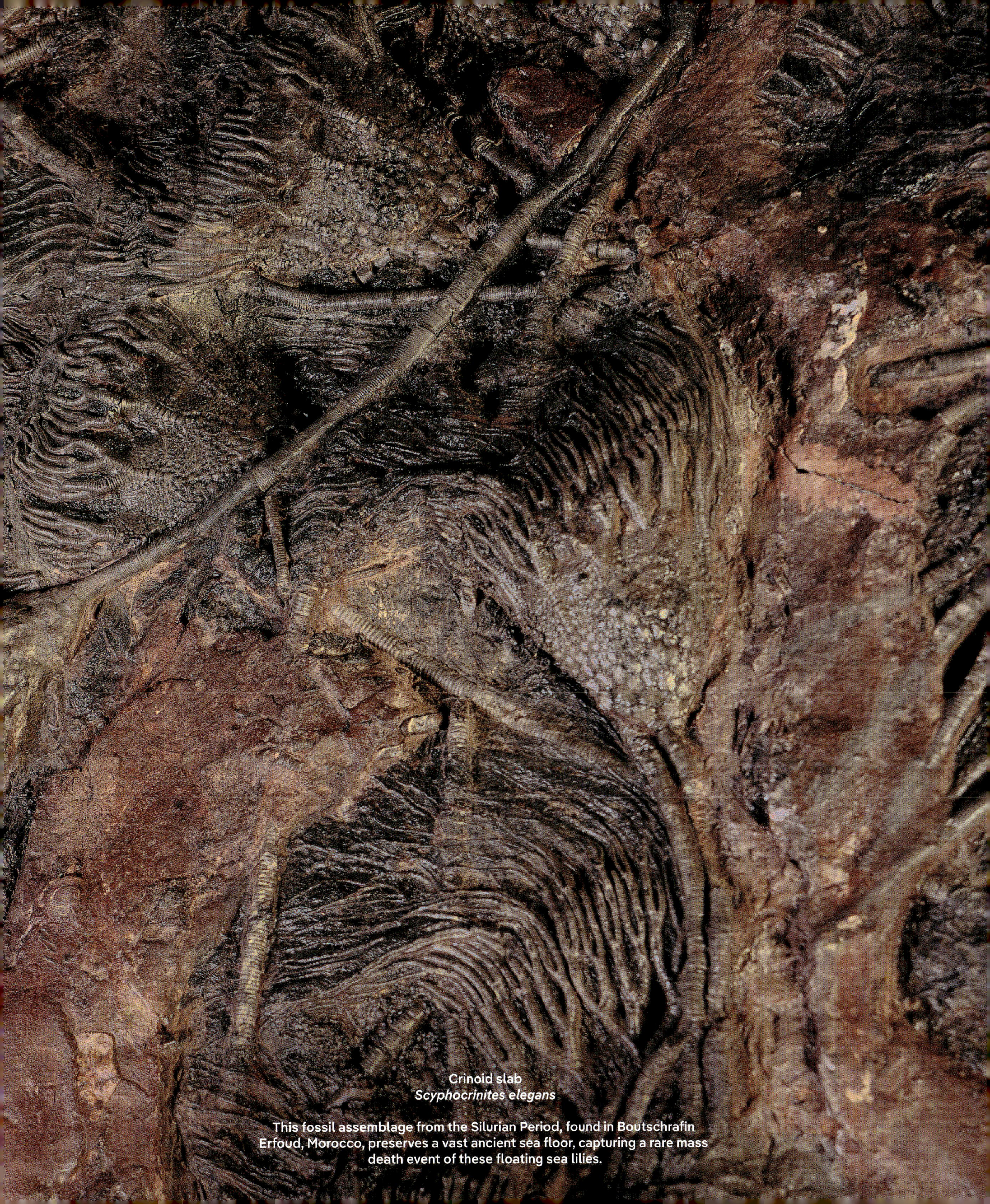

Crinoid slab
Scyphocrinites elegans

This fossil assemblage from the Silurian Period, found in Boutschrafin Erfoud, Morocco, preserves a vast ancient sea floor, capturing a rare mass death event of these floating sea lilies.

***Hidden Treasures* publication would not have been possible without the support of**

Department of Culture
and Tourism - Abu Dhabi

HE Mohamed Khalifa Al Mubarak
Chairman

HE Saood Abdulaziz Mohamed Al Hosani
Undersecretary

Rita Aoun-Abdo
Executive Director, Culture Sector

Dr Peter C. Kjærgaard
Natural History Museum Abu Dhabi, Director

Curatorial
Maryam Salman Alhaddad
Mark Jonathan Beech
Luca Belucci
Rebecca Desjardins
Victoria Egerton
Ludovic Ferrière
Virginia Harvey
Brigitte Howarth
Hanneke Jacobs
Sebastian Kirchhof
Phillip Lars Manning

Editor
Mona Farag

Collections, Conservation
Hanna AlHassen
André Appelt
Corina Berners
Lodevico Jr. Gazmen
Lawrence Salu Fernandes
Irene Cazalla Manceras
Judith McAlester
Pari Tamilveeran
Aline Toukhmanian
Stanislav Vinnichenko

Photography, Digital
Dmytro Kotelnikov
Marc McAlester
Live Streamlines, Construction of lapse photography

Skira editore

Head of Business Development
Edoardo Ghizzoni

Managing Editor and Head of Production
Manuela Calandra

Design
Giovanni Fanelli

Layout
Filippo Anceschi

Editorial Coordination
Claudia Podio

First published in November 2025 by
Department of Culture and Tourism - Abu Dhabi
PO Box 94000
Abu Dhabi, United Arab Emirates
Tel.: +971 2 444 0444; Fax: +971 2 599 5720
abudhabiculture.ae

Skira editore srl
via Agnello, 18
20121 Milano
Italy
skira-arte.com

ISBN: 978-88-572-5506-4 (DCT)
ISBN: 978-88-572-5455-5 (Skira)

This book has been printed by RGM Printing on FSC®-certified paper

FSC www.fsc.org MIX Packaging | Supporting responsible forestry FSC® C102233

Distributed in USA, Canada, Central & South America by
ARTBOOK | D.A.P., 75 Broad Street, Suite 630, New York, NY 10004, USA.
Distributed elsewhere in the world by Thames and Hudson Ltd., 181A High Holborn, London WC1V 7QX, United Kingdom.